WHY IT'S NOT ALL ROCKET SCIENCE

WHY IT'S NOT ALL ROCKET SCIENCE

SCIENTIFIC THEORIES AND EXPERIMENTS EXPLAINED

 Thames & Hudson

Robert Cave

CONTENTS

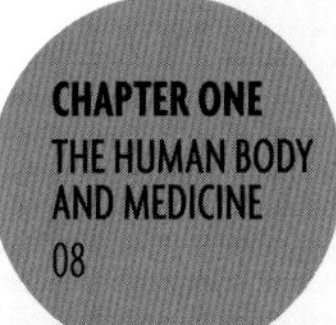

CHAPTER THREE
SOCIETY
96

CHAPTER FOUR
THE PLANET
138

CHAPTER FIVE
THE UNIVERSE
178

INTRODUCTION

ROBERT CAVE

'Eureka!'—or 'I've found it!'—is an exclamation associated with discovery, elucidation and scientific enquiry that dates back more than 2,200 years to Greek mathematician and scientist Archimedes. According to legend, he was asked to establish whether the gold crown of Hieron II, King of Syracuse, contained the same amount of precious metal that had been given to the goldsmith for the purpose of creating it. Crucially, Archimedes was not allowed to damage the crown as he tested its composition. The scientist was contemplating how to accomplish this task while relaxing in his bathtub when the answer came to him. He observed the displacement of water as he moved around in the bath and realized that he could place the crown in water and then measure the amount of water it displaced. If he compared this measurement to the amount of water displaced by the quantity of gold that Hieron II had given to the crown maker, he would be able to prove experimentally whether or not the crown was made solely from the gold provided. This revelation is reported to have been so profound that Archimedes jumped out of his bath and exclaimed 'Eureka!' while running naked down the street.

The truth and practicality of this particular moment has been questioned by historians and scientists alike, but the scientific principle behind the experiment—that gold has a different density to other metals—is accurate, as is the rush of excitement that has often accompanied scientific discoveries and the successful execution of experiments to test them.

This book attempts to capture some of the excitement and strangeness of experimental science. Covering a period of more than 300 years, *Why It's Not All Rocket Science* highlights one hundred experiments and discoveries—unusual, controversial, accidental or pure genius—that have made a significant impact on scientific developments, either at the time or since. Some experiments, such as Arthur Eddington's proof of Einstein's theory of general relativity, have been hugely consequential, whereas others—Thomas Parnell's pitch drop experiment, for example—are more of an exercise in critical thought and demonstrate that experiments can last much longer than a human lifespan. Some of the scientists' work might seem strange, such as Justin Schmidt submitting to various insect bites in order to rate pain sensation, but Percy Spencer's experiments with early radar equipment led to a new commercial everyday product: the microwave oven.

Like the universe itself, the frontiers of the process of science keep expanding. As such, this book cannot be a definitive or comprehensive guide to the most significant experiments conducted. Instead, it is a selection of interesting, odd and far-reaching experiments. Some are massive in scale, such as the confirmation of the Higgs boson by teams of scientists at the Large Hadron Collider, whereas Thomas Young's double-slit experiment, for example, is small and simple enough to conduct yourself. Self-experimentation also has an important part to play, but this is certainly not advisable to try at home. In addition, there are at least three systems of classification—Linnaean taxonomy, Mendeleev's periodic table and the standard model of elementary particles—none of which can be described as complete. However, their incompleteness serves as a reminder that science is not set in stone, but a dynamic ongoing process of discovery based on experimental evidence. As such the focus of the book is on experimental science, the moments beyond the mathematics that demonstrate the theories.

Chapter one—The Human Body and Medicine—examines experiments that have given us a greater understanding of the mechanisms of our bodies. Chapter two—Psychology and Behaviour—covers the classic conditioning work of Ivan Pavlov and continues into experiments on conformity and obedience. Chapter three—Society—explores an assortment of studies and demonstrations that have impacted on wider society. Chapter four—The Planet—looks at experiments that relate to the place we call home, from discerning Earth's age to mapping its last unexplored regions. The final chapter—The Universe—analyses the science of space travel and space exploration as well as the science of the fundamental components of all matter.

Above all, *Why It's Not All Rocket Science* is a testament to the work of scientists and their pursuit of knowledge through experimental evidence. They are driven in their labours by the motto of the Royal Society: *Nullius in verba*, meaning 'take nobody's word for it'. Their experiments did not all proceed from correct assumptions, nor did they always result in answers that were fully accurate. However, they have led to an understanding of how the world works that is incrementally progressively less wrong. Scientific experimentation is not perfect, and may never produce all the answers, but it does give us the tools and the method.

GUIDE TO SYMBOLS

Explains why the experiment is an important scientific work.

Outlines the scientist's other work and achievements.

Locates the experiment in its historic and scientific context.

Unattributed quotes are by the scientist featured.

Provides additional incidental information.

Briefly describes the accompanying image.

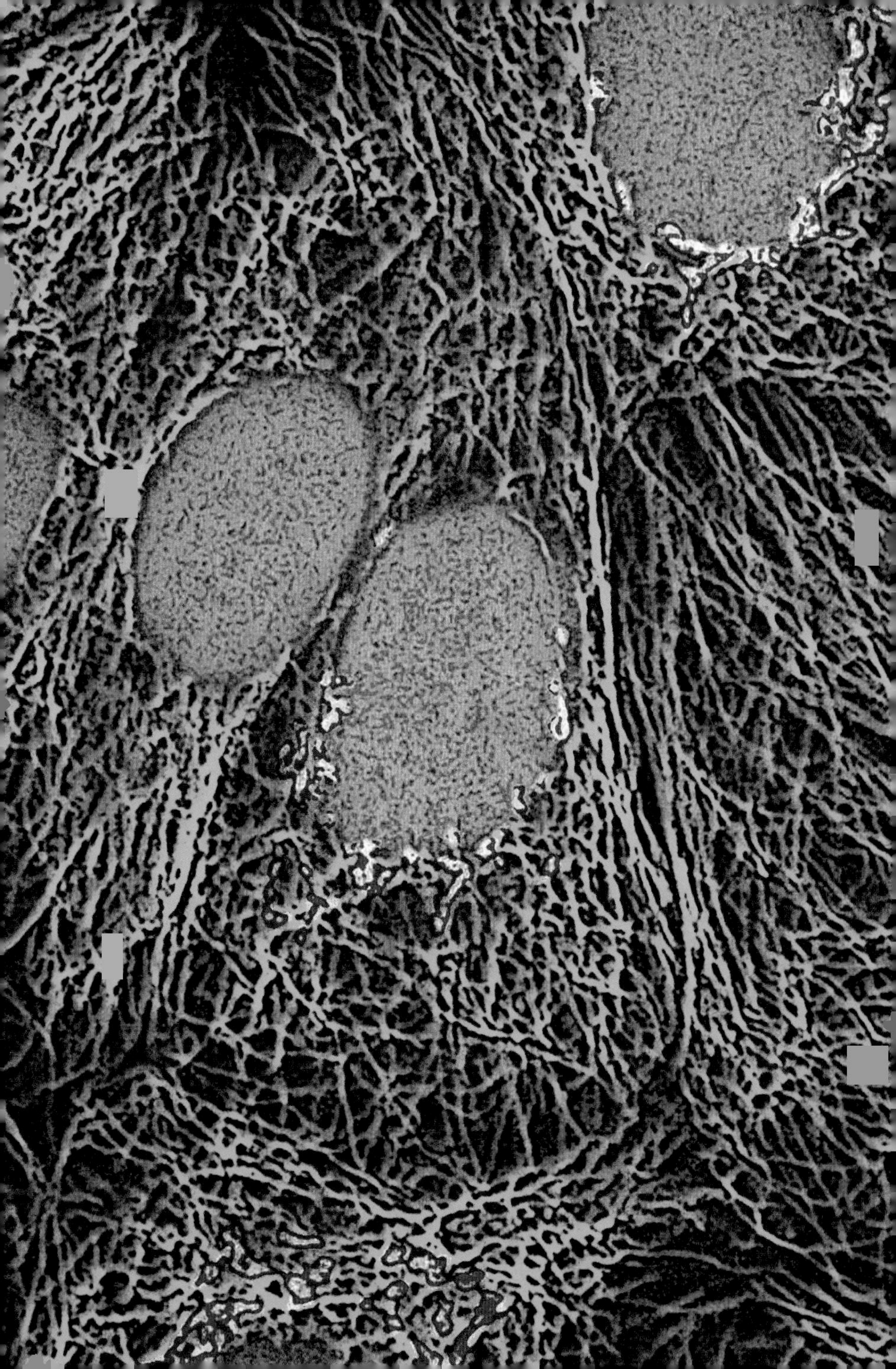

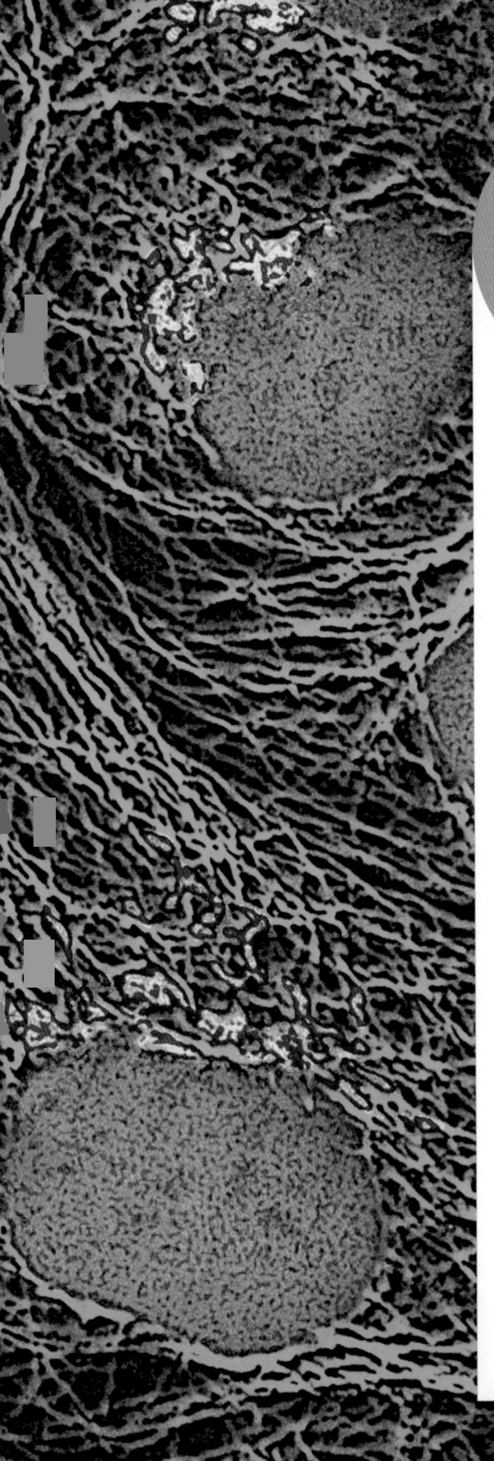

CHAPTER ONE
THE HUMAN BODY AND MEDICINE

Throughout much of history, the workings of the body have been something of a mystery to doctors. However, through fortuitous accidents, bold self-experimentation and scientific genius, medical practitioners have established ever-growing insight into the functioning of organic biology and the human body. Often, their experiments focused on discovering the root cause of a disease, as was the case with Joseph Goldberger and pellagra, but at other times they were designed to prove the efficacy of a potential treatment, such as Alexander Fleming's work with penicillin, or to develop a diagnostic technique, for example, Werner Forssmann's heart catheterization experiment. Fundamentally, the purpose of their research was to gain a better understanding of how the body worked, both in sickness and in health.

‹ HeLa Cells (see p. 34)

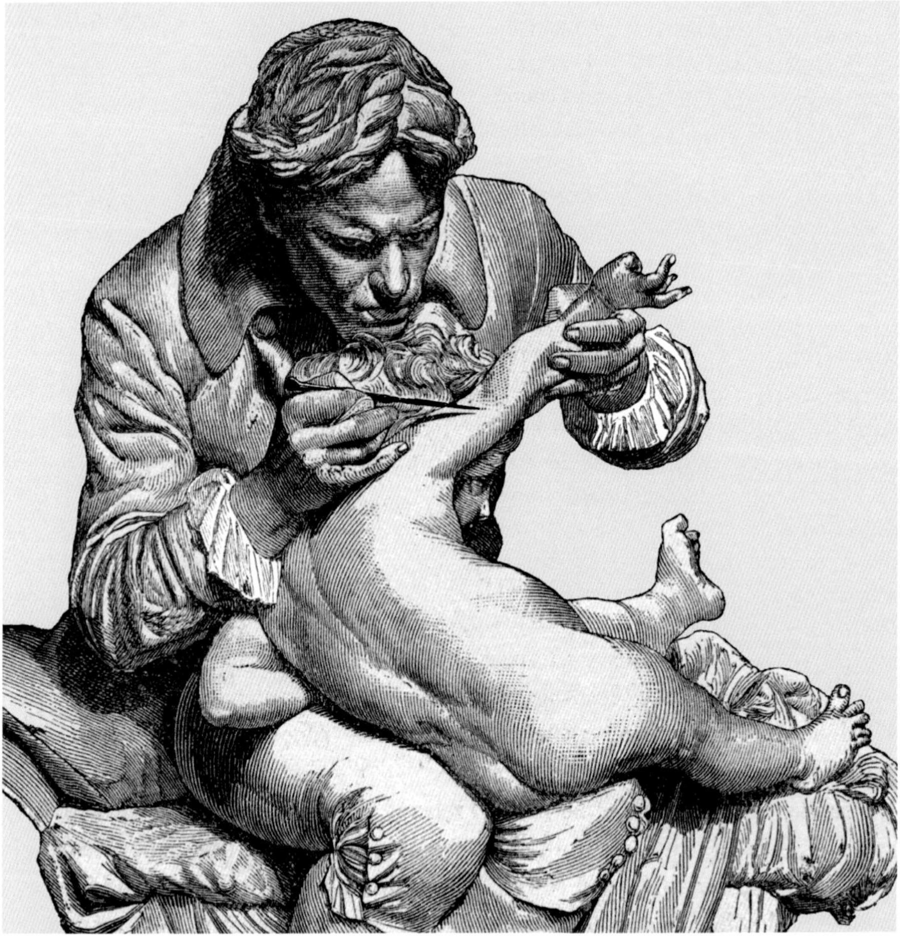

Although Jenner's conclusions met with initial resistance, doctors soon began to use his technique, which was termed 'vaccination' from the Latin root *vaccinus* meaning 'relating to the cow'. Vaccination proved to be safer and more reliable than variolation, and by the early 19th century its use had spread across Europe and reached the United States. In 1840 variolation was banned in the United Kingdom and by 1853 vaccination had become compulsory. The fight against smallpox continued until the virus was declared extinct outside of laboratories in 1980.

From *Heroes of Britain in Peace and War* (*c.* 1897), this illustration depicts Jenner vaccinating his own son.

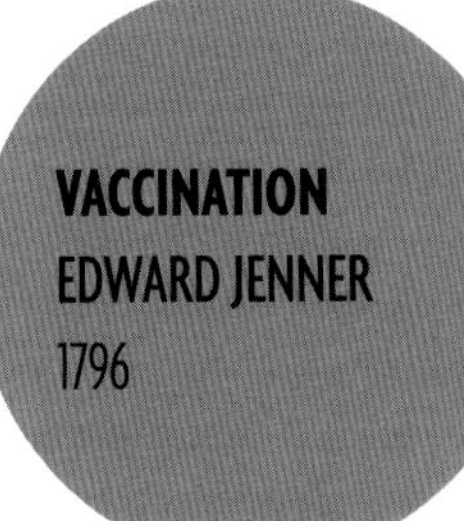

Edward Jenner was serving as an apprentice to a country surgeon in Gloucestershire, England, when he first heard accounts of milkmaids who had suffered from cowpox yet boasted that they would never contract smallpox. After qualifying as a doctor, Jenner began to investigate these claims, and he built up a set of case studies of cowpox sufferers who had shown resilience to smallpox. This evidence led Jenner to postulate that exposure to the milder cowpox infection could protect against smallpox. He assembled a large collection of case studies, including that of a local dairymaid called Sarah Nelmes, and set out to inoculate a healthy eight-year-old boy named James Phipps. On 14 May 1796, Jenner placed pus that he had collected from Nelmes's cowpox lesions into two small cuts, each about 1 centimetre (0.4 in.) long, that he had made on Phipps's arm. The boy suffered from a slight fever seven days after the inoculation, but otherwise he made a complete recovery. He was subsequently exposed to 'matter' from the pustule of a smallpox sufferer, but no disease developed and he was declared to be immune. Jenner repeated the experiments over the following years before presenting his findings to the Royal Society in 1798.

Jenner (1749–1823) was the son of a vicar. Orphaned at a young age, he was raised by his sisters and variolated with smallpox while at school. At the age of twenty-one, he moved to London and expanded his field of studies to include the natural sciences. He helped to classify samples brought back by Captain James Cook from his first voyage to the Pacific Ocean, and then returned to his native village of Berkeley in 1772 to practise medicine and pursue his cowpox investigations. In 1784 Jenner built and launched his own hydrogen balloon, and in 1788 he published a study into the nesting habits of the cuckoo, the latter of which won him election to the Royal Society.

By the late 18th century, smallpox was a well-known disfiguring disease, but it was also known that those who recovered from smallpox gained some immunity to it. A level of protection could be induced through variolation—the deliberate controlled infection of those who had never had smallpox—but this process was not always successful and could lead to a potentially fatal outbreak of the disease.

Several slight punctures and incisions were made on both his arms, and the [smallpox] matter was carefully inserted, but no disease followed.

YELLOW FEVER
STUBBINS FFIRTH
1802–03

In early 19th-century Philadelphia, Pennsylvania, outbreaks of yellow, or 'malignant', fever were a regular occurrence, but the city's inhabitants had little clue as to the cause of the disease and how to avoid becoming infected. After studying a number of outbreaks, US medical student Stubbins Ffirth thought he could prove that yellow fever was not contagious, and set out his evidence for such a conclusion in his doctoral dissertation. In *A Treatise on Malignant Fever; with an Attempt to Prove its Non-Contagious Non-Malignant Nature* (1804), Ffirth noted that in past cases the infected patients had often failed to infect those with whom they lived in close proximity. He conducted a series of experiments that attempted to induce infection, first in cats and dogs and then on himself, mainly through exposure to the bloody black vomit of severely infected yellow fever sufferers. Ffirth dropped vomit into his own self-inflicted wounds and on his eyeball; he cooked the vomit and inhaled the fumes, fashioning the remains into tablets that he consumed; finally, he drank the vomit, both diluted and freshly obtained from a yellow fever patient. He also exposed himself to other bodily fluids from the afflicted. None of this led to Ffirth contracting yellow fever, and he therefore concluded that the disease was not contagious.

Ffirth (1784–1820) began his studies at the University of Pennsylvania in Philadelphia, a city that had been severely affected by the yellow fever epidemic in 1793. He noted that the disease was rife during the summer months but almost disappeared in winter. During a second epidemic in 1802–03, he conducted his series of yellow fever experiments and presented his findings in 1804, when he was serving as the house surgeon to the Philadelphia Dispensary.

Why then, oh! Ye . . . physicians of Columbia, will you not lay aside your prejudices, investigate nature and judge for yourselves?

This illustration (1820) is from a work by French physicians Etienne Pariset and André Mazet; it shows the tongue of a yellow fever victim at various stages of the infection.

In Ffirth's day, the fear of infection associated with the outbreak of disease affected both commerce and the care of the infected. It was not unusual for patients to be shunned by their frightened families. Although Ffirth did not establish the true cause of the disease, his findings on yellow fever and the dangers of panic are part of the US surgeon general's library.

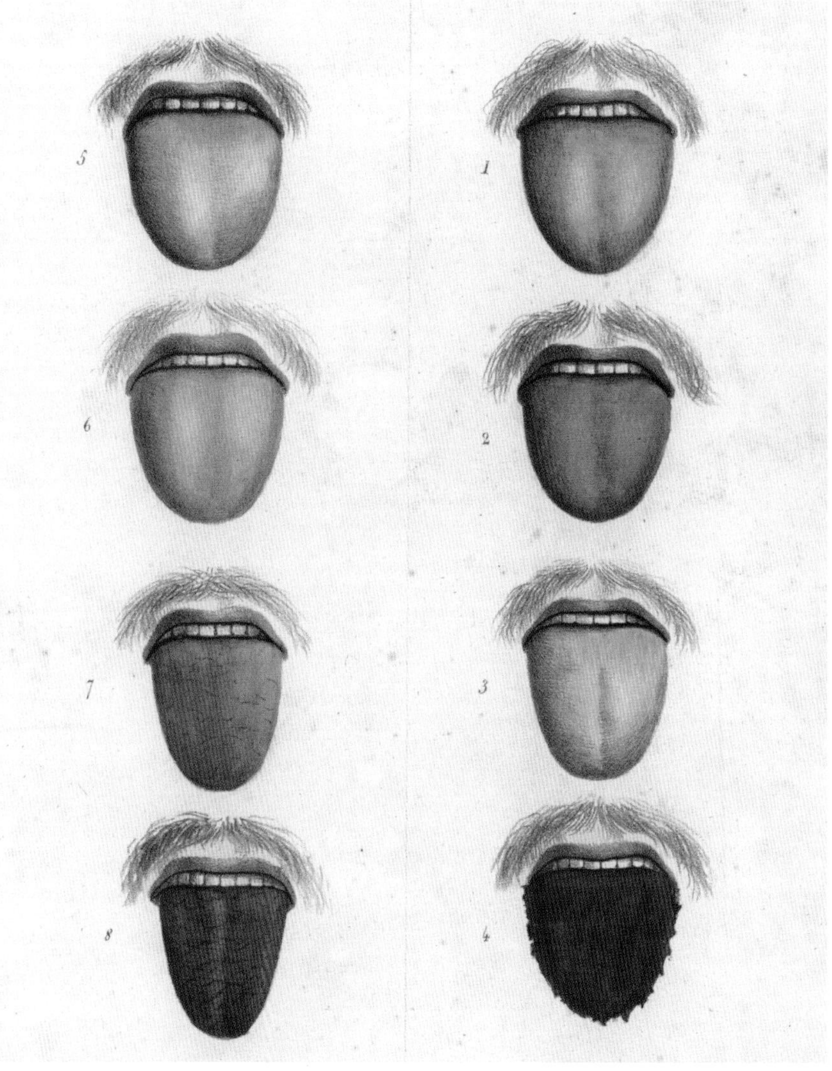

Ffirth's conclusion that yellow fever could not be transmitted through close contact was correct, but he failed to establish how humans came to be infected with the disease. However, Carlos Finlay and Walter Reed later identified that yellow fever was infectious and mosquitoes were the disease vector for transmission. We now know it to be a blood-borne virus.

Beaumont (1785–1853) studied medicine before joining the US Army and becoming a surgeon, stationed at Fort Mackinac close to the US/Canadian border. Today, he is regarded as the 'father of gastric physiology' and is often credited as the United States' first great medical scientist.

DIGESTION
WILLIAM BEAUMONT
1825–33

In August 1825, William Beaumont began a series of experiments on digestion on a Canadian fur trapper named Alexis St. Martin, who had been accidentally shot in the chest. Against all expectation, St. Martin survived the incident but developed a permanent fistula, or hole, under his left breast, thereby allowing direct access to his stomach. Beaumont was astonished by the opportunity for study that the fistula provided, declaring: 'I can look directly into the cavity of the stomach, observe its motion and almost see the process of digestion.' He attached various pieces of food to silk strings and dangled them through the fistula into St. Martin's stomach, withdrawing them at intervals to assess their digestion. This caused the patient some discomfort, so Beaumont administered medicine directly through the fistula. Next, he ascertained the temperature of St. Martin's stomach by placing a thermometer through the fistula. After recording a temperature of 37.7°C (100°F), Beaumont then used a rubber tube to siphon off about 30 millilitres (1 fl oz) of gastric juice into a vial. He placed a small piece of salt beef into the liquid, sealed the vial and heated it to the constant temperature of his previous observations. Over the following nine hours, Beaumont observed the clear gastric juices slowly break down the meat until all that was left was a cloudy soup of fluid.

In the early 19th century, the complex workings of the living human body remained something of a mystery among medics. Anatomists dissected dead bodies, which revealed little of the processes of life, such as digestion.

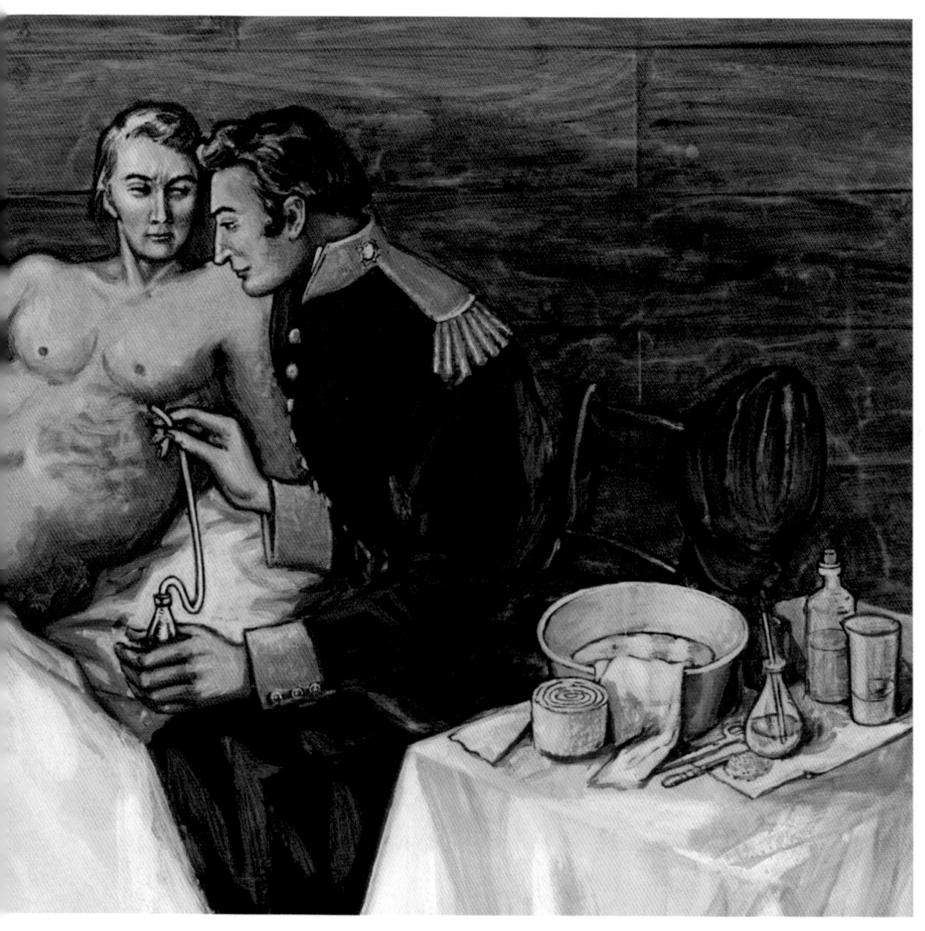

Over a period of nine years, Beaumont conducted more than 230 experiments on St. Martin as he made the most of this unique opportunity for scientific study. In 1833 he published an account of them titled *Experiments and Observations on the Gastric Juice and the Physiology of Digestion*. Although the doctor's medical ethics and treatment of St. Martin as an experimental resource rather than as a human are sometimes questioned, the results were invaluable in establishing the natural temperature of the stomach and that food did not need to be chewed in order for the stomach to digest it.

Early depictions of Beaumont's experiments, such as this, tend to emphasize his military background while minimizing the discomfort of the patient.

1

3

Although Darwin was yet to develop his theory of evolution through natural selection when he observed the close relationships of the Galápagos birds, they serve as a prominent example of how speciation can arise in animals that diversify over generations in reproductive isolation. He went on to investigate human evolution in *The Descent of Man, and Selection in Relation to Sex* (1871).

Ornithologist John Gould's drawings show the differences in beak shape among the various finches Darwin presented.

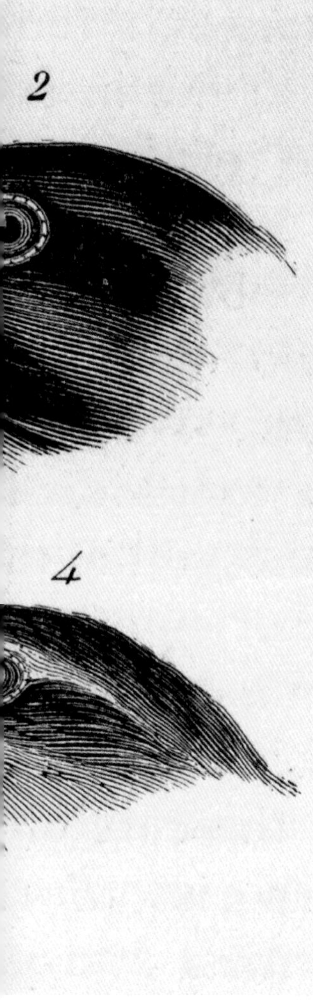

Some of the most important scientific breakthroughs are made as the result of investigations rather than experiments. Gould's observations prompted Darwin to re-examine the birds that he had collected on the voyage and to reconsider the implications of their diversity with regard to their close relatives.

DARWIN'S FINCHES
CHARLES DARWIN
1830s

The second voyage of HMS *Beagle* (1831–36) was intended to survey various points in South America, but the ship's captain, Robert FitzRoy, wanted a geology expert on board, too. British naturalist Charles Darwin landed the role and took to the voyage with the intention of collecting interesting animal and plant specimens. With his servant Syms Covington, he shot and preserved bird specimens from various locations on four islands of the Galápagos, but he did not keep detailed notes on where each bird had been found. Nor did he initially pay much attention to the species of bird that would later collectively bear his name. It was only when he delivered his collection of birds to the Royal Geographical Society in 1837 that ornithologist John Gould examined the specimens and recognized that they represented at least twelve new species of closely related finches.

The term 'Darwin's finches' was popularized by David Lack in his 1947 book, which aimed to demonstrate the process of evolution through natural selection.

Darwin (1809–82) showed a passion for natural history from a young age, growing to become the father of the life sciences. Ubiquitous for his theory of evolution, *On the Origin of Species* (1859), Darwin used supporting observations made by his peer, and natural selection advocate, Alfred Russel Wallace.

Mendel (1822–84) came from a poor Austrian family and joined a religious order of Augustinian Friars so that they would pay for his education. In 1854 he was permitted to conduct a series of hybridization experiments at the monastery in Brünn. Over a two-year period, he studied thirty-four varieties of peas and presented his findings to the Natural Science Society in 1865. A year later he published his paper 'Experiments on Plant Hybrids' to a lukewarm reception. In 1868 Mendel was promoted to abbot and abandoned much of his scientific research due to the increased workload.

This historic artwork depicts three generations of Mendel's pea experiment, a tiny fraction of those he actually bred in the grounds of his monastery.

In 1936 English statistician Ronald Fisher calculated that Mendel's results were too accurate—'too good to be true'—and accused those close to Mendel of editing his data.

HEREDITY
GREGOR MENDEL
1850s

From his family's background in farming, botanist Gregor Mendel knew that *Pisum sativum* garden peas would be good plants to use in his experiments on heredity. They had at least seven easily observable characteristics, he could control their pollination and they grew quickly, which would allow him to study multiple generations with relative ease. Mendel identified that each characteristic, such as pea colour, had two expressions, or phenotypes: peas were either green or yellow. For each characteristic, he grew two pure lineages of plants: one for each phenotype (A and B, opposite). Over time, purity was established via monitoring, so that one plant group yielded only green peas and the other only yellow. Mendel then cross-pollinated the two lines and noticed that the resulting peas were all yellow (C). This allowed him to infer that the yellow pea phenotype was dominant. He then grew these yellow peas into the next generation of plants through self-pollination. The resultant peas from this generation were either yellow or green in a ratio of three yellow to one green (D), prompting the scientist to describe the green phenotype as recessive. After observing later generations, Mendel concluded that the expression of characteristics was controlled by a then-unnamed inherited factor in the seed. Today, we know this factor as a gene.

It had long been observed that animals and plants pass down their characteristics through the generations, but no general law governing the distribution of these characteristics had been discovered by the mid 19th century. Focusing on pea plants and their hybrids, Mendel investigated the framework that governed inheritance.

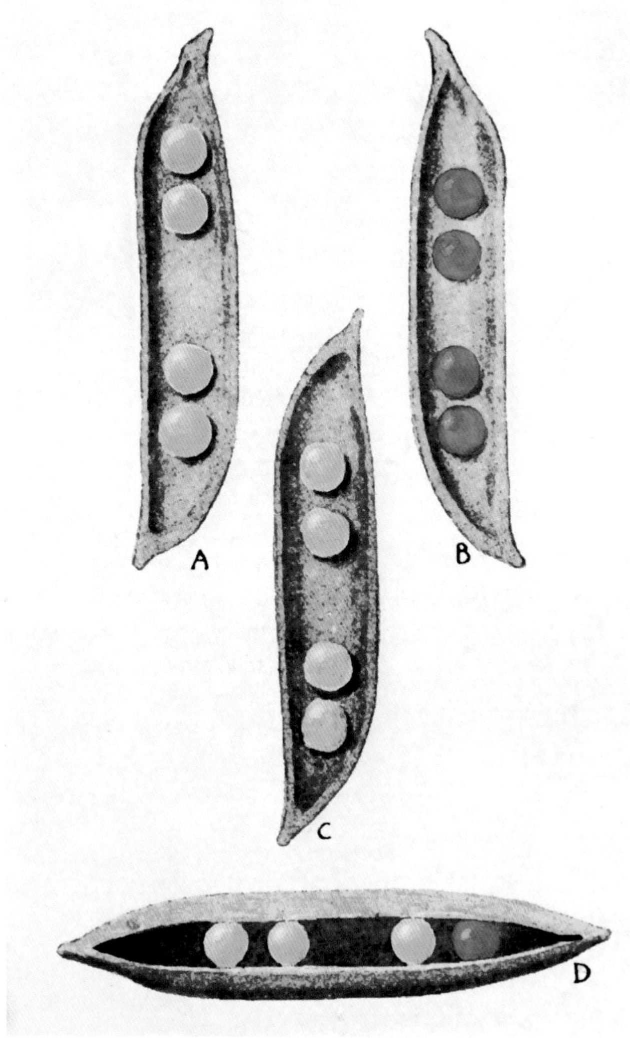

?

Mendel established the foundation of genetics, but the importance of his work was not fully understood during his lifetime. In 1900 Dutch botanist Hugo de Vries rediscovered Mendel's heredity laws, and today these experimental studies are widely recognized as having established the rules by which hereditary characteristics are transmitted through generations.

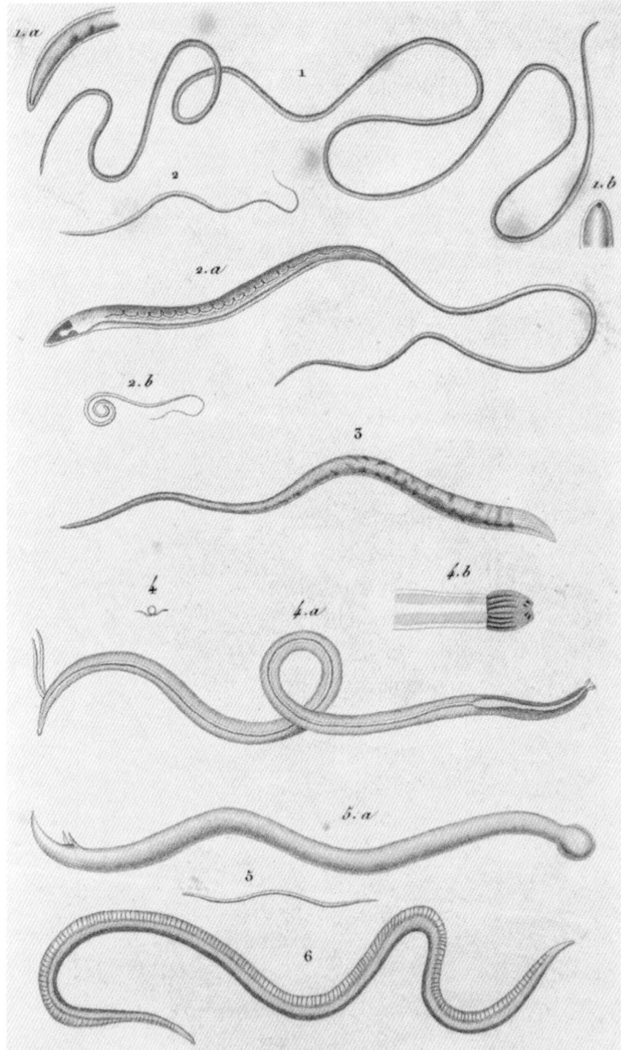

Grassi proved that *A. lumbricoides* could enter the human food chain directly if its eggs were consumed, and that it did not need an intermediate host. Self-experimenter Shimesu Koino subsequently traced its movements within the human body as the larvae moved from the intestines to the liver, into the bloodstream, through to the lungs, up the trachea and back to the small intestine as an egg-laying adult.

ROUNDWORM LIFE CYCLE

GIOVANNI GRASSI

1878–79

Italian physician Giovanni Grassi's investigation of the life cycle of the *Ascaris lumbricoides* roundworm involved a bizarre self-experiment. It began in 1878 when he conducted an autopsy on a man whose stomach and intestines were full of adult *A. lumbricoides* roundworm and their eggs. Grassi harvested some of the eggs and stored them in excrement to preserve them. As the subject of the experiment, he had to establish that he was not already infected with roundworm, and so carefully examined his own faeces for a period of ten months for any trace of parasitic infection. After finding nothing remarkable in his stools, Grassi was satisfied that he was free from roundworm. On 30 August 1879, he directly consumed some of the eggs that he had harvested from the corpse. After a period of twenty-two days, Grassi started to feel some abdominal discomfort. Upon examining his faeces, he discovered a number of fresh roundworm eggs, which indicated that the creatures were now living inside him. He then began taking herbal anti-worming medicine to drive the roundworms out of his system. Grassi repeated this experiment in 1887, when he fed *A. lumbricoides* eggs to a seven-year-old boy. Over three months, the boy evacuated 143 worms of lengths up to 23 centimetres (9 in.).

By the end of the 19th century, it was known that a variety of worms were able to parasitize humans, but they were often only discovered after death through autopsy. It was not well understood how roundworms got into the body nor what could be done to prevent further infections. Grassi was willing to do whatever it took to investigate the *A. lumbricoides* life cycle.

Although Grassi (1854–1925) studied medicine at the University of Pavia, he retained a fascination for zoology, and particularly parasites, throughout his life. He is probably best-known for demonstrating the link between *Anopheles claviger* mosquitoes and malaria in humans, but his work was overlooked by the Nobel Prize committee, who instead handed the recognition to British army doctor Ronald Ross for his earlier observations that mosquitoes were the transmitting disease vector behind malaria in birds.

Naturalist Georges Cuvier depicted various parasitic worms in his book *The Animal Kingdom* (1817), but unlike Grassi, he never went to extraordinary lengths to study their life cycle.

Grassi would have needed an especially strong stomach for his experiment: the worms are well known for their strong and unpleasant smell, which they retain even if carefully cleaned.

Little is known about medical examiner von Niezabitowski outside of his writings. However, these indicate that the enquiry into corpse fauna was conducted at the request of Leon Jan Wachholz, a Polish pioneer of forensic medicine.

How long do carrion insects need . . . to completely destroy a human body?

Von Niezabitowski established that the same fly larvae and beetle species that feast on dead animals will freely devour the corpses of humans, too. He noted that it took only two weeks for corpse fauna to strip every last piece of flesh from the bones of cadavers left out in the summer. This evidence helped pave the way for the modern field of forensic entomology, which is today used to deduce the time of death from the species and development of the insects and larvae found on dead bodies.

Von Niezabitowski described a number of flies and beetles as corpse fauna, including the Silphidae family of beetles.

CORPSE FAUNA

EDUARD RITTER VON NIEZABITOWSKI

1899

Like any good scientist, Eduard Ritter von Niezabitowski approached his investigation into the fauna associated with the decomposition of corpses in a systematic way, first reviewing all the existing medical literature on the topic, which he described as 'sparse'. He concluded that he would need to obtain experimental evidence for himself and secured the dead bodies of a variety of small creatures—cat, fox, rat, mole and calf—as well as several corpses of stillborn babies. These he arranged in a local vegetable garden and in some of the more secluded areas around the Jagiellonian University, Krakow (now part of Poland), before letting nature take its course. Von Niezabitowski returned to the corpses on a daily basis to observe exactly what types of creature invaded the bodies during decomposition and which were the first to arrive. He soon noticed a swarm of flies in the vicinity of the corpses. A large number of these had laid their eggs in the soft tissue, with the common greenbottle (*Lucilia caesar*) being the most prolific. Beetles arrived later in the decomposition process, around a week after the cadavers were initially deposited. Von Niezabitowski repeated his experiment over the course of a year in order to assess whether the seasons affected the fauna involved. Although he noted some seasonal variation in the number and variety of certain insects, his main finding was that corpse fauna was most active during the summer months and that bodies were devoured most rapidly at that time.

Founded in 1364 by Casimir III the Great, Jagiellonian University is named after Poland's Jagiellonian dynasty of rulers.

Scientists had little formal knowledge of what happened to bodies after death. French entomologist Jean Pierre Mégnin had published a book—*La Faune des Cadavres* (1894)—but questions remained as to how long it would take before all the soft tissue on an exposed corpse was consumed.

PROTOPATHIC AND EPICRITIC PAIN

HENRY HEAD

1903–07

Early in his medical career, Henry Head developed an interest in how patients who had suffered from peripheral nerve injury often recovered touch sensation later on. He realized that, as a trained doctor with experience of the human somatosensory system, he was the ideal patient to articulate and chart how sensation returns after nerve damage. Consequently, he elected to make himself the subject of an experiment. On 25 April 1903, James Sherren operated on Head's left forearm, first dividing two cutaneous nerves. Taking care to avoid harming any muscle tissue, Sherren then severed the peripheral branches of the radial nerve and immediately sutured them back together. Over the following four years, Head made regular visits to the home of psychiatrist and neurologist William Halse Rivers, who examined and recorded first where Head had lost sensation and then progressively when and where and to what degree it returned. Initially, Head could not discern anything touching his skin, but the area retained 'deep sensibility, tactile and painful response to pressure'. Rivers tested Head's hand for sensitivity to a variety of stimuli, including pins, cotton wool and objects of various temperatures, as he charted how fine touch and temperature sensation returned. He also adapted the experiment and blindfolded Head because he noticed his friend's tendency to allow visual and aural cues to affect his perception of touch.

This photograph was taken thirty-one days after the operation; it shows a triangular area on Head's wrist that was sensitive to fine touch, but not pinpricks.

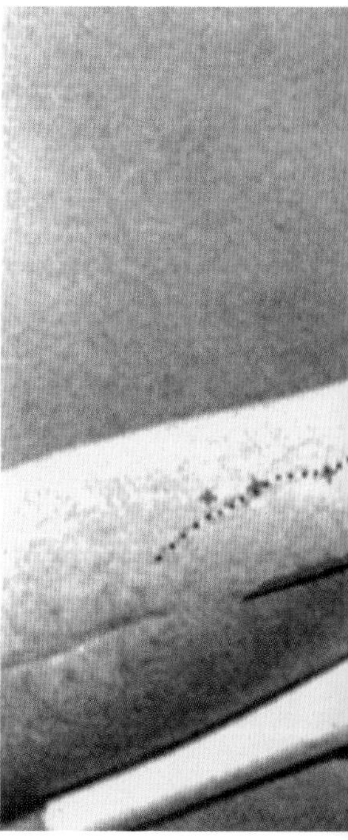

By the end of the 19th century, the theory that pain was an unbearable intensification of ordinary sensation—espoused by German neurologist Alfred Goldscheider and others—had largely been set aside in favour of the notion that pain was separate from ordinary touch sensation. Head was eager to probe further into the workings of the human nerve system.

The glans of the penis typically does not respond to epicritic sensation, and Rivers performed comparative tests on this part of Head's anatomy, too.

Born in London into a wealthy family, Head (1861–1940) developed a fascination with medicine at an early age and studied at Cambridge before commencing his medical career in 1890. A gifted poet and doctor, his passion for more accurate knowledge drove his involvement in the pain experiment.

The areas of the hand innervated by the radial nerve were completely incapable of detecting sensation . . .

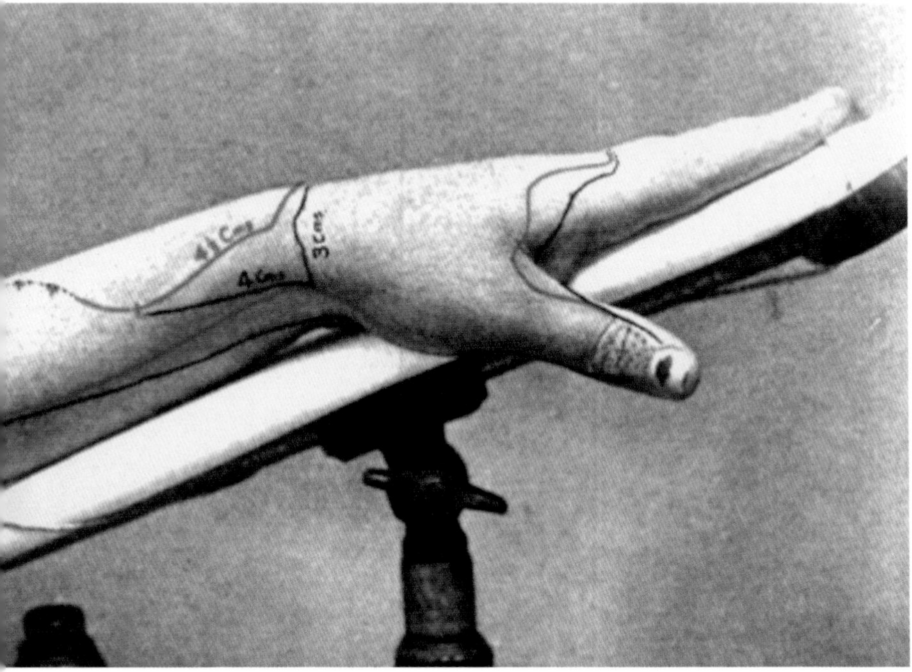

As a result of their experiment, Head and Rivers divided the regenerated sensation into two fundamental areas. Their conclusion defined these two types of sensation and established the distinction between them. Protocritic sensibility could detect broad changes in temperature and pressure without the ability to localize where the sensation originated. Epicritic pain was more discriminatory and responsible for fine tactile sensation and the ability to discern a greater range of temperature. Exteroceptive, proprioceptive and interoceptive sensations were also suggested as further classifications.

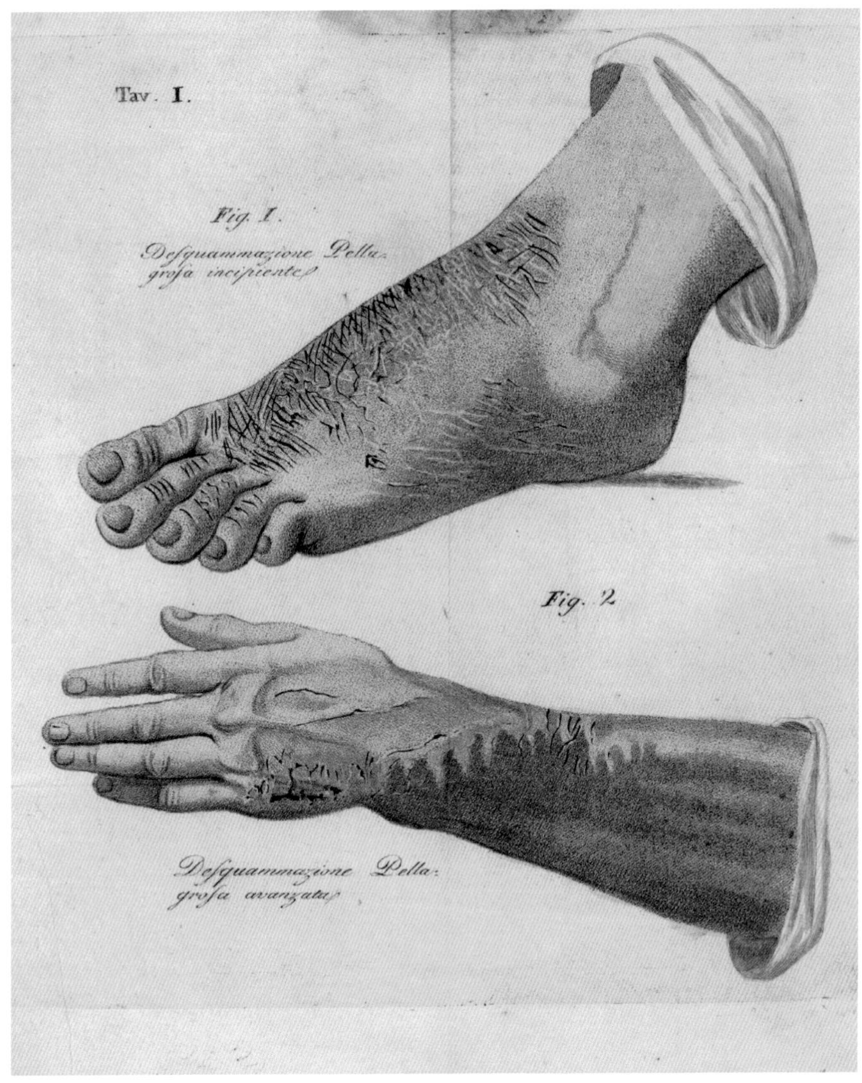

Pellagra is a vitamin deficiency that can cause skin lesions, diarrhoea, dementia and death, but it was once thought to be an infectious disease. Goldberger's experiments demonstrated that this was not the case, yet political resistance to his theories remained. It was not until 1937 that Conrad Elvehjem used Goldberger's work to isolate niacin (vitamin B3) and identified its role in preventing pellagra.

During two years of research and study, physician Joseph Goldberger noticed that while prisoners and asylum inmates developed pellagra, their nurses and wardens did not. This indicated that pellagra was not infectious, but was caused by poor diet. However, politicians and the medical establishment remained unconvinced, so Goldberger resorted to a more flamboyant method of demonstrating his theory. In 1915 he induced pellagra in six prison volunteers by limiting their diet. Next, he wanted to prove that pellagra would not result from exposing healthy people to those suffering from the condition. In 1916 he collected samples of blood and various bodily effluvia—urine, nasal mucous, faeces and scabby skin scales—from pellagra victims. He then recruited sixteen volunteers, including his own wife, and exposed them to this material. Initially, Goldberger and his assistant, Dr George Wheeler, had injected one another with 5 to 6 cubic centimetres (0.3–0.35 cu in.) of blood from a pellagra sufferer with no ill effect. Goldberger then took the effluvia and mixed it with wheat flour and cracker crumbs to fashion it into tablets he consumed with his volunteers at events dubbed 'filth parties'. Although some of the volunteers experienced nausea and diarrhoea (they were eating faeces), none of them developed pellagra.

Diagnoses of pellagra, a potentially lethal condition, spiked in the southern United States in the early 20th century. In 1914 Goldberger was charged with the task of establishing the cause of the disease. His studies of prisoners, orphans and asylum inmates led him to theorize that it was caused by a limited diet.

Born in Hungary, Goldberger (1874–1929) emigrated to the United States with his family in 1883. He graduated with a Doctor of Medicine in 1895 and set up his own private practice in Pennsylvania before joining the US Public Health Service in 1899. There, he conducted health inspections on new immigrants in New York. He later studied epidemiology and the effects of parasites in disease transmission in the southern United States and Mexico.

If pellagra be a communicable disease, why should there be this exemption of the nurses and attendants?

This illustration of pellagra in the hand and foot dates back to 1814 and was drawn by Italian physician Vincenzo Chiarugi. Although pellagra was particularly prevalent in the southern parts of the United States, it became known around the world in the early 20th century.

DISCOVERY OF PENICILLIN

ALEXANDER FLEMING

1928

Throughout his medical career, Alexander Fleming developed a reputation for keeping an untidy laboratory, allowing the petri dishes of bacterial cultures to stack up. However, this somewhat slovenly habit paid off when he noticed that a *Staphylococcus* culture that had been exposed to the air had become contaminated by an unknown mould that appeared to be killing the bacteria. With help, and through observing the mould structure, Fleming was able to identify the mould as a member of the *Penicillium* genus of fungus. He followed up the discovery with a series of experiments, investigating the best medium in which to grow the mould—in this case, nutrient broth—and whether the antibiotic effect was reduced by filtering the mould (it was not). Fleming tested the filtrate, which he dubbed penicillin, against various types of bacteria to ascertain which were susceptible to its antibiotic effect. He also subjected penicillin to different temperature levels and dilutions to see how these affected its antibacterial properties. In order to assess its toxicity, he injected it into rabbits and mice and poured it onto a man's eyeball.

Fleming (1881–1955) was born in Scotland into a large farming family, but was encouraged by his older brother to study medicine. At medical school, he joined the rifle club and was such a good shot that the captain suggested he get a job in the school's bacteriology department (as a way of keeping Fleming in the team after graduation). He served in the Royal Army Medical Corps during World War I, where he became aware of the need for better anti-bacterial agents to reduce the number of deaths. His work was recognized in 1945 when he was jointly awarded the Nobel Prize in Physiology or Medicine with Chain and Florey.

During World War I, Fleming witnessed at first hand that the antiseptics used to combat sepsis and bacterial infection in wounded soldiers were inadequate. They frequently weakened the patient's immune response and failed to attack anaerobic bacteria in deep wounds. Fleming's immediate recommendation that infected tissue should be cut away was not well-received. After the war, he returned to his studies, little suspecting he would find a more effective alternative to traditional antiseptics.

This mounted specimen of the *Penicillium* bacteria originally belonged to Alexander Fleming. It helped to kick off a whole revolution in the development of antibiotics.

Fleming discovered the antibacterial enzyme lysozyme when he added his own nasal mucus to weak bacterial cultures.

Fleming's results showed that penicillin inhibited and killed a range of bacteria at dilutions of up to 1 in 800. It was more effective than many antiseptics and, best of all, it was not toxic. Biochemist Ernest Chain and pathologist Howard Florey were later inspired by Fleming's results to persuade drug companies to mass manufacture penicillin. By the end of World War II, the Allied powers were using the drug extensively to reduce the number of deaths and amputations from bacterial infections.

When I woke up just after dawn on September 28 1928, I certainly didn't plan to revolutionize all medicine by discovering the world's first antibiotic, or bacteria killer.

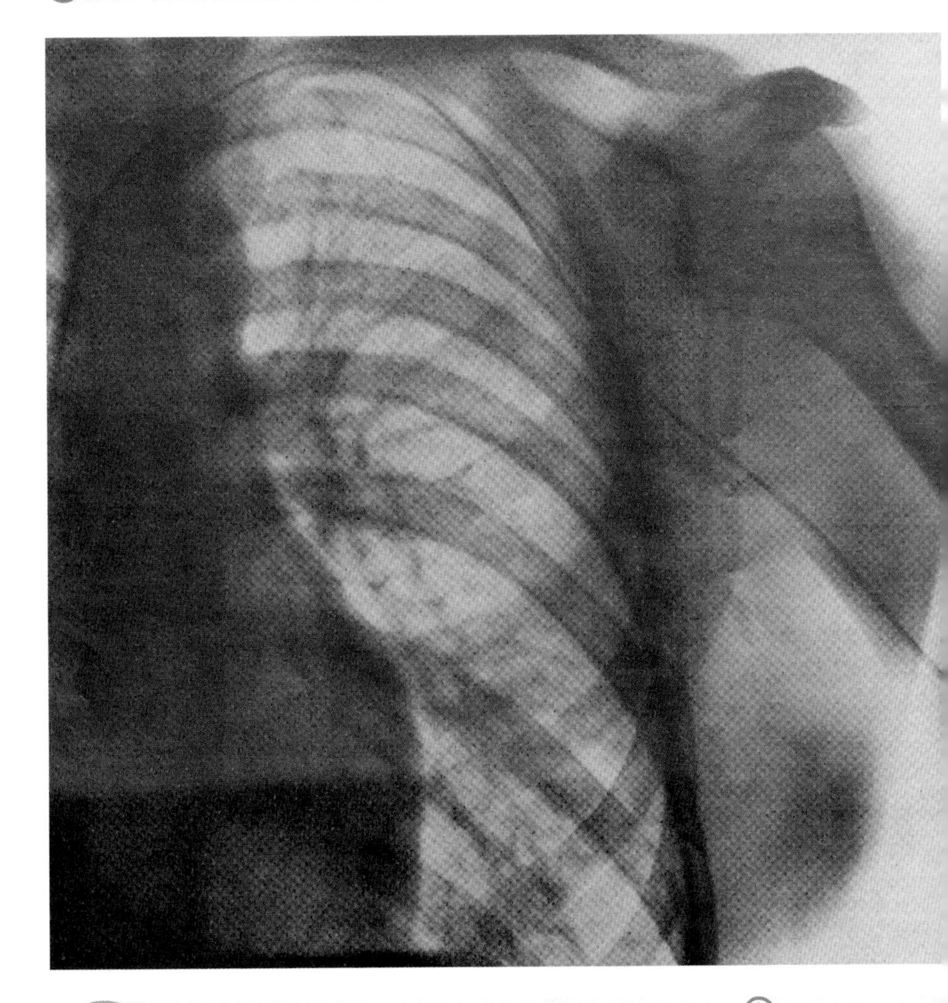

Forssmann correctly believed that his idea could help doctors improve their diagnoses of cardiac complaints. However, it was only after World War II that his work was championed by US doctors André Cournand and Dickinson Richards. Together, the trio was awarded the Nobel Prize in Physiology or Medicine in 1956 'for their discoveries concerning heart catheterization and pathological changes in the circulatory system'. Subsequently, heart catheterization has been put to therapeutic uses, with some doctors using the technique to remove blockages.

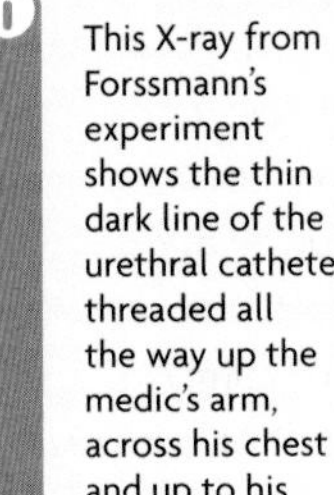

This X-ray from Forssmann's experiment shows the thin dark line of the urethral catheter threaded all the way up the medic's arm, across his chest and up to his heart.

CARDIAC CATHETERIZATION

WERNER FORSSMANN

1929

In the early 20th century, it was widely believed that it was unacceptably risky to attempt any surgery on the heart. However, this did not dissuade Werner Forssmann—a young medical intern—from hypothesizing that it was possible to catheterize the heart by threading a long narrow tube through a vein in the arm. When Forssmann told his bosses of his plans to test this theory on a patient, they refused him permission to proceed, suggesting he first conduct his experiment on animals. Forssmann ignored them and convinced the nurse in charge of surgical instruments of the importance of his experiment. She agreed to help on the condition that she would be the subject. Forssmann acceded, but then tricked the nurse and tied her to the operating table while he anaesthetized his own left arm. He then opened his antecubital vein close to the elbow and began to insert a small metal tube intended for urethral catheterization. After working the tube up to his shoulder, Forssmann went to the X-ray department where he used a fluoroscope—a device that produces moving X-ray images—to monitor the tube as he threaded it into his right auricle, near the atrium of his heart. He then captured a chest X-ray as proof of his success.

Forssmann (1904–79) was born into a wealthy Berlin family that fell on harder times after his father was killed in World War I. Encouraged by his family and teachers, Forssmann studied medicine and passed his final exams at the University of Berlin in 1928. Although he was noted for his medical skill, his unorthodox cardiac catheterization experiment brought him notoriety, which prompted him to switch his specialization from cardiology to urology. Forssmann served as a military surgeon in World War II and had largely given up hope that cardiac catheterization would become widely adopted when he learned of his Nobel Prize win. Ironically, he died of heart failure.

Prior to Forssmann's self-experiment, much of the early work on the study of the heart had been conducted on animals. This was largely due to ethical or moral concerns about performing interventions on the hearts of human patients. Fortunately, Forssmann was not put off investigating his own heart, and wrote up the results in his paper 'Die Sondierung des rechten Herzens' (The Probing of the Right Heart, 1929).

Forssmann was a member of the National Socialist German Workers Party from 1932 until the end of World War II, which affected his ability to practise medicine for several years after the war ended.

After the Mammoth Cave experiment, Kleitman remained convinced that sleep required much greater study. He went on to publish the first edition of his landmark book *Sleep and Wakefulness* in 1939, which surveyed the existing literature on the subject. He subsequently updated this seminal text in 1963.

SLEEP/WAKE CYCLES
NATHANIEL KLEITMAN
1938

As part of his investigation into sleep/wake cycles, sleep researcher Nathaniel Kleitman explored extending and shortening the 'days' of volunteers in laboratory conditions, but daylight and the noise of the working day intruded. He needed somewhere more secluded for his experiment. With no natural light and a constant temperature of 12°C (54°F), the Mammoth Cave, Kentucky, seemed to be the ideal venue for testing sleep patterns. It was 20 metres (65 ft) wide and 8 metres (26 ft) high and was located 40 metres (130 ft) underground. Unfortunately, the cave had high humidity, and rats. Undeterred, Kleitman furnished the cave and ordered meals from a local hotel. On 4 June 1938, the researcher and his assistant, Bruce Richardson, started living twenty-eight-hour days, split into nineteen hours of activity (ten hours of work, nine of leisure) and nine hours of sleep. They took their oral temperatures every two hours during the 'day' and every four hours during the 'night' to chart against their normal fluctuations of temperature. After thirty-two days, the pair emerged from the cave to significant press attention.

The temporary cave home was well-furnished in order to keep Kleitman and Richardson busy and engaged as they attempted to stave off sleep.

Kleitman (1895–1999) was born in what is now Moldova into a Jewish family and was driven from his homeland to the United States by pogroms and World War I. From early on in his academic career, he studied the subject of sleep, and he continued to do so well into retirement, participating in sleep studies while in his nineties. He was also involved in an investigation into the alcoholic intoxication of beer after the end of Prohibition.

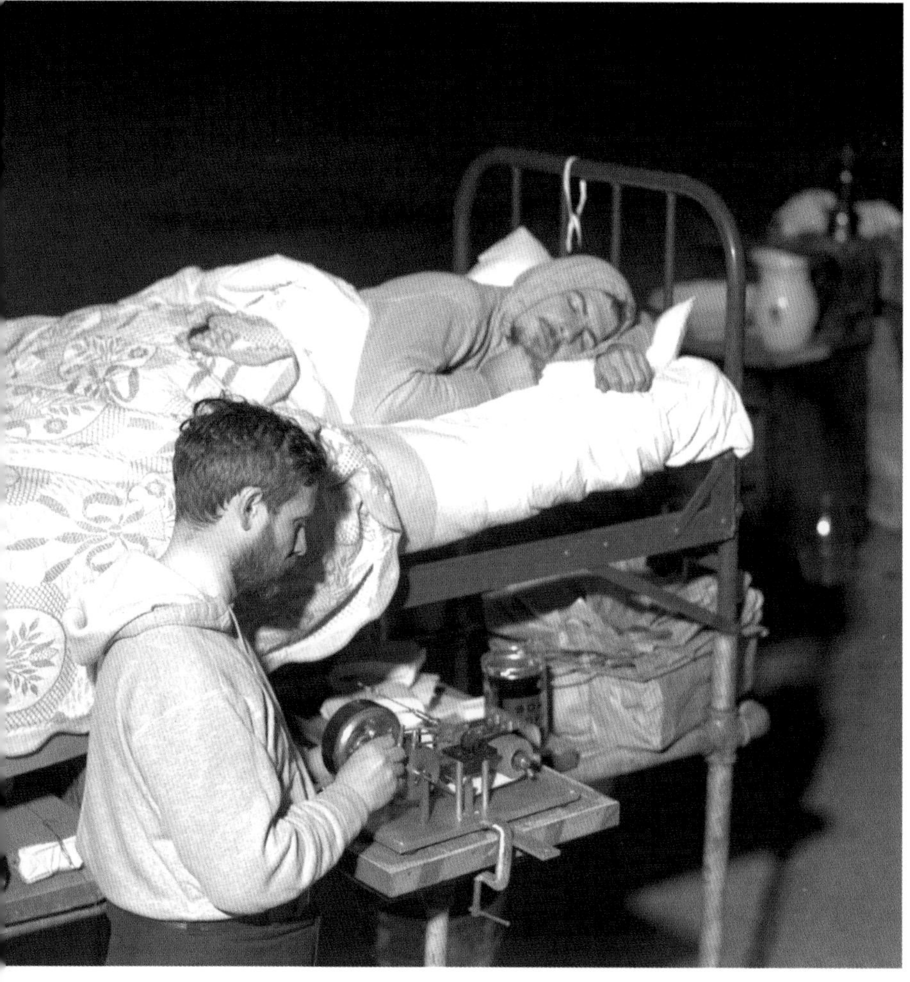

The results of this exercise were inconclusive: Richardson seemed to have adapted to twenty-eight-hour living, whereas Kleitman had not. Despite this, Kleitman subsequently established himself as the father of modern sleep research. His standout achievement was the discovery, with Eugene Aserinsky, of rapid eye movement (REM) sleep in 1953 and its link to dreaming.

Although it was a highly unusual field of study, Kleitman's research received early backing from the company behind Ovaltine, the well-known malty bedtime drink.

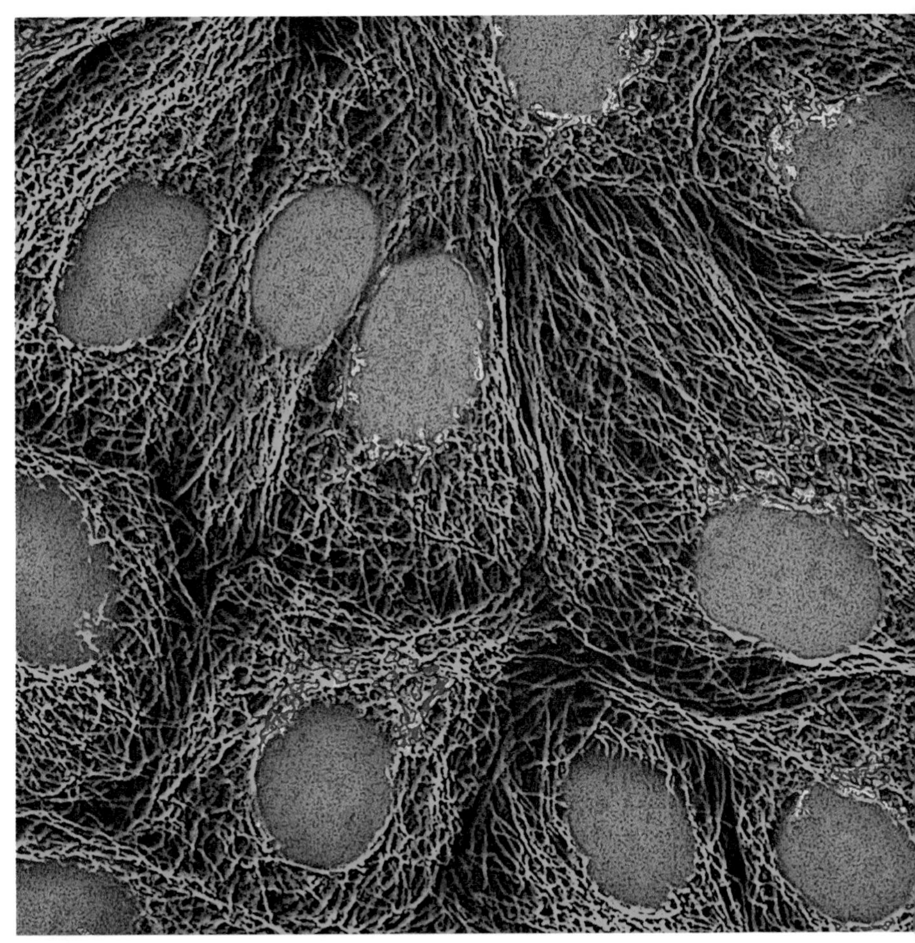

?

Although it is debatable whether discarded material, such as cells, should be used for research without the patient's consent, there is no doubt that Gey's actions were of great benefit to the future of medical research. In fact, it is difficult to quantify exactly how important HeLa cells have become to biomedical science. As of September 2015, more than 85,000 papers had been written citing the use of these cells for 'research into cancer, AIDS, the effects of radiation and toxic substances, gene mapping and many other scientific pursuits'.

i

Here, the nucleus containing the HeLa genes is coloured blue, the protein modifying Golgi bodies is orange and the cytoskeleton that facilitates intracellular transport is green.

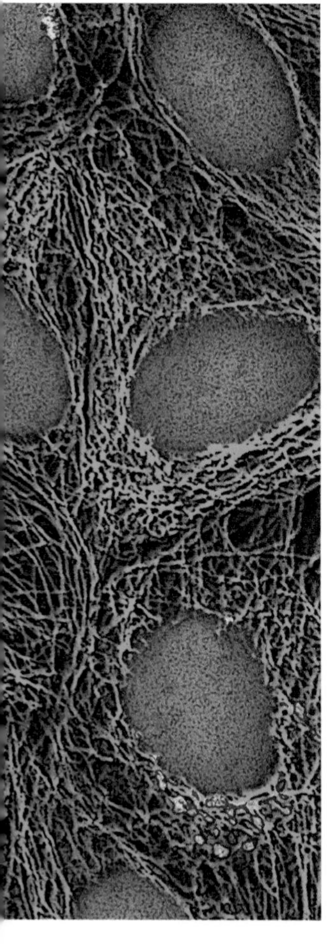

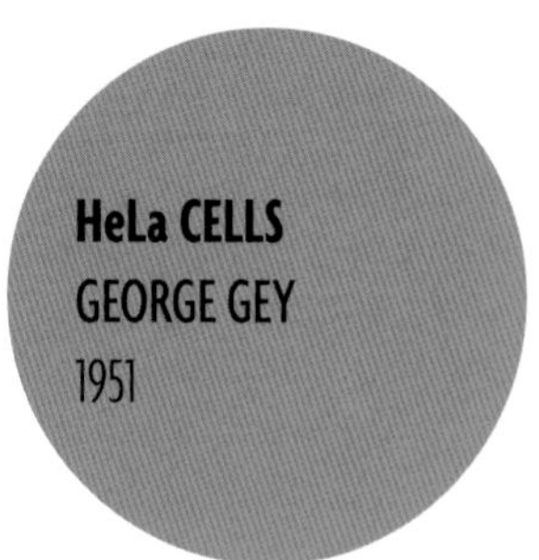

HeLa CELLS
GEORGE GEY
1951

In January 1951, a poor, thirty-year-old black woman named Henrietta Lacks went to the John Hopkins Hospital in Baltimore, Maryland. Lacks did not know it at the time, but she was suffering from an aggressive form of cervical cancer that would kill her within the year. The doctor who examined her took a sample of the cancer for a biopsy and then, without her knowledge or consent, passed some of the cells to physician George Gey, who had requested that all samples taken from patients be brought to him to assist in his research. Gey then attempted to grow more of the cells in his laboratory. Human cells had only been observed to divide a finite number of times before they died, but to his astonishment he discovered that not only did some of these cells grow quickly, but also that they continued to multiply. These cells, which he dubbed HeLa cells after Henrietta Lacks, were the first human-derived cells that were effectively immortal to be propagated in a laboratory. Gey recognized how useful such cells could be for medical tests and immediately began distributing them to anyone who requested them for the benefit of science. By 1952 HeLa cells were already being used by US scientist Jonas Salk to help test the first polio vaccine.

In the mid 20th century, scientists could grow human-derived cells in a laboratory, but they had a limited lifespan. This severely limited the scale and scope of biomedical research.

Gey (1899–1970) was born in Pittsburgh, Pennsylvania, to German immigrant parents. He secured a degree in science before gaining his doctorate in medicine from John Hopkins University in 1933. There, he worked on the problem of growing human cells in laboratory conditions. Discovering HeLa cells was his ultimate breakthrough, and he kept the cells' precise origins a secret up to his death.

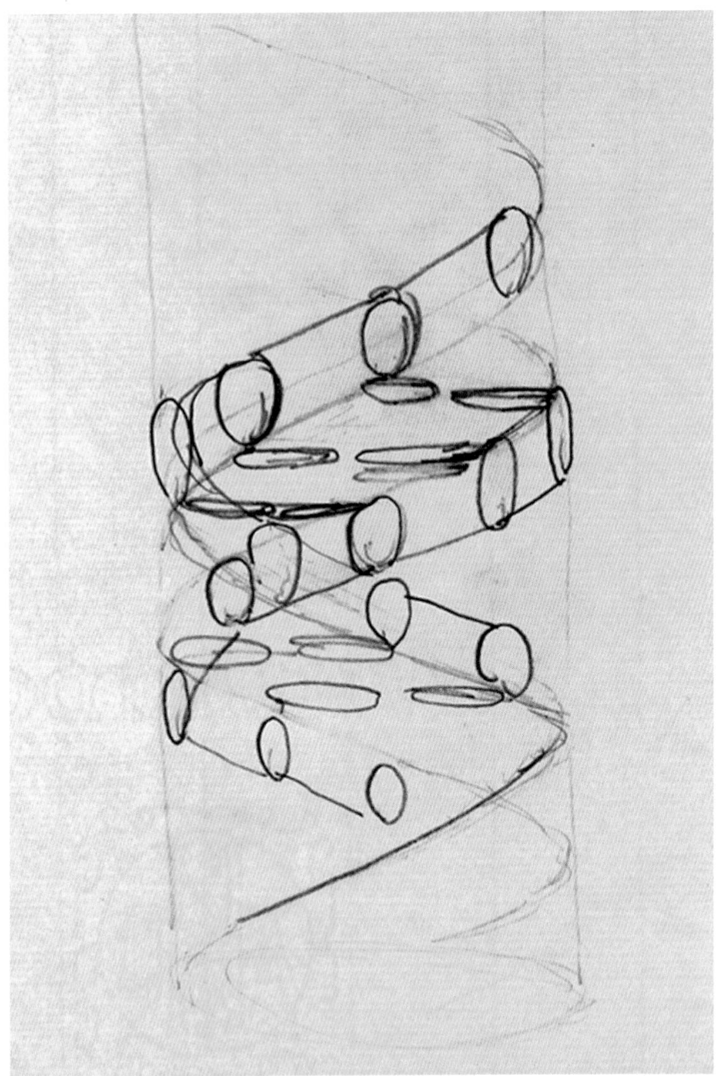

?

The realization that DNA has a double helix structure revolutionized the field of genetics and provided valuable insights into the mechanism of base pairing through which heritable and other genetic information is stored. It is one of the most important discoveries of the 20th century and paved the way for later landmarks, such as the human genome project, leading to a broader understanding of how life works.

STRUCTURE OF DNA
FRANKLIN, CRICK AND WATSON
1951–53

In 1951 British chemist Rosalind Franklin was brought to King's College London on a three-year research scholarship and tasked with joining the burgeoning field of DNA research. She focused on investigating the structure of DNA salts using crystallography techniques, an area in which she was a respected expert. Franklin worked at improving her equipment and her samples in order to produce better images of what had been identified as two different forms of DNA, termed A-DNA and B-DNA. The two forms varied in their level of water content, with B-DNA being the 'wetter' of the two. Franklin was still investigating the structure when her colleague, Maurice Wilkins, showed one of her crystallographic diffraction photographs of B-DNA to Francis Crick and James Watson without her knowledge. The latter pair had been working on models of DNA based on what was known of its chemical composition. The photograph prompted Crick and Watson to focus on a double helix design in their subsequent modelling work and led to the realization that DNA formed a double helix structure. Their paper proposing this architecture was published in 1953 in the science journal *Nature*. It featured Franklin's crystallography work on DNA, complete with the X-ray diffraction image that had galvanized Watson and Crick's discovery.

Although DNA was first isolated as far back as 1869 by Swiss biologist Friedrich Miescher, its significance in the role of heredity was not understood. Oswald Avery's theory of this role was proposed in 1944 and experimentally demonstrated eight years later. These events increased the interest in DNA and its structure, prompting more scientists to focus on its study.

Crick (1916–2004) and Watson (1928–) shared with Wilkins the 1962 Nobel Prize in Physiology or Medicine for 'their discoveries concerning the molecular structure of nucleic acids and its significance for information transfer in living material'. Sadly, Franklin (1920–58) had passed away and was not eligible to share the Nobel Prize for her experimental work in the discovery of the structure of DNA. However, she has gained recognition for her contribution in recent years.

Although they were colleagues at King's College, Wilkins and Franklin apparently had a very difficult relationship and did not get on at all well.

Crick's original sketch from 1953 shows both the twin helical structure of DNA and the arrangement of base pairs that form each rung on the DNA ladder.

Demikhov (1916–98) was a very innovative surgeon, but the committee of the Soviet Ministry of Health pronounced his experimental work to be unethical. Despite this, he went on to achieve international recognition, and in 1989 he received the Pioneer Award from the International Society for Heart and Lung Transplantation in acknowledgment of his work.

HEAD TRANSPLANT
VLADIMIR DEMIKHOV
1954

After years of organ-focused transplant experiments, Soviet surgeon Vladimir Demikhov made a name for himself in 1954 by transplanting the head and upper limbs of a small dog onto the neck of a much larger dog. News of this experiment slowly filtered around the world, and in 1959 journalists from the US publication *Life* magazine travelled to Russia. They watched Demikhov's team as they cut into a small dog that was sleeping under general anaesthetic. An assistant made the first incision, rolling back the little dog's skin, before Demikhov stepped in. Using his scalpel to expose the blood vessels, and a needle and thread to draw them closed with knots, he cut the small dog's upper body, heart and lungs free from its hindquarters. Another assistant cut into a shaved area on the large dog's neck, exposing its heart and veins. Demikhov then connected the main blood vessels of the small dog to the corresponding vessels of the larger one, adroitly splicing them together with a surgical stapler of his own design, until the small dog was fully attached to the large dog's circulation. The small dog's heart and lungs were then cut away and its torso, complete with head and fore paws, was stitched to the large dog's neck. The whole operation took only three-and-a-half hours.

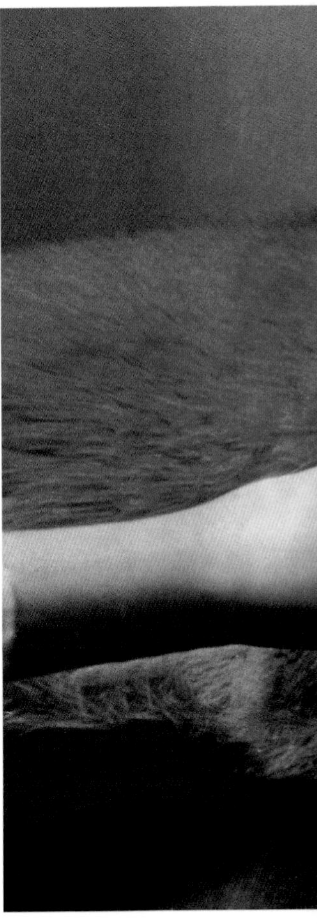

Alexis Carrel won the Nobel Prize in Physiology or Medicine in 1912 for pioneering the sewing together of blood vessels, but it was not until the 1930s that surgeons began to develop the techniques required for successful organ transplants.

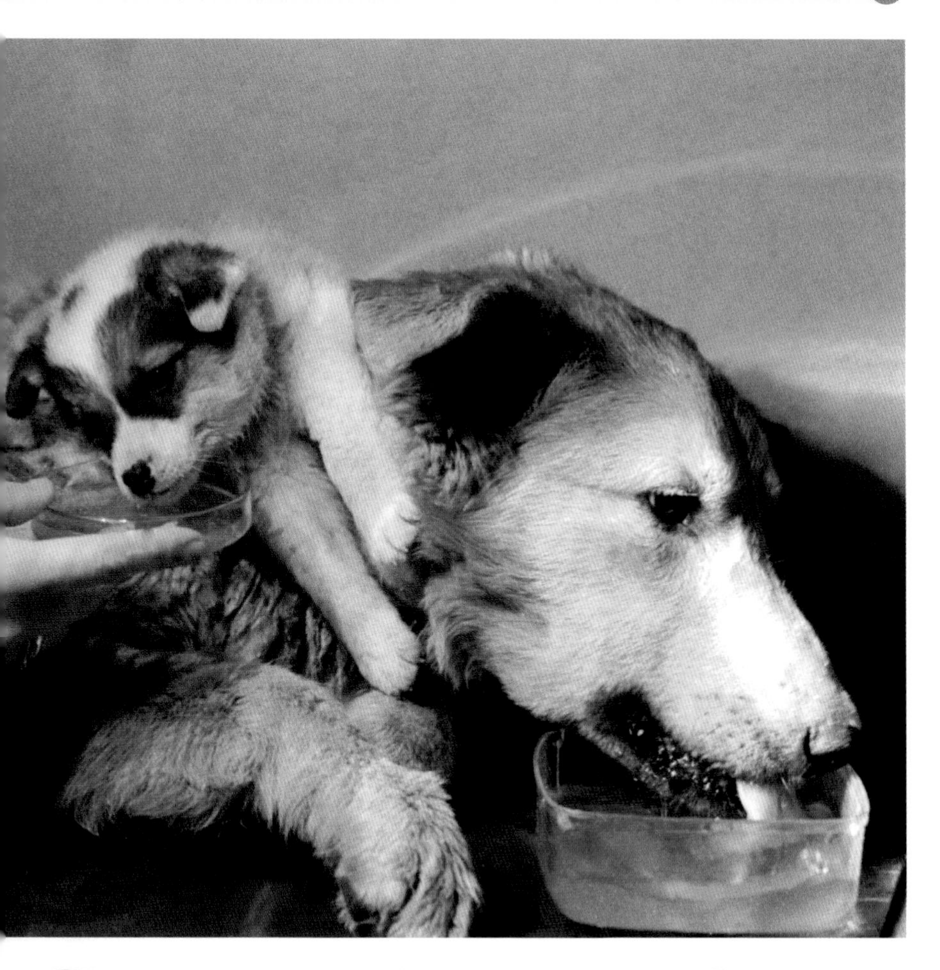

Demikhov's various transplant experiments on dogs helped him to pioneer an artificial heart pump and also the first successful heart, lung and liver transplants on animals. His work was little known in the West until a monograph of his experiments was translated into English in 1962, which influenced Christiaan Barnard, the surgeon who performed the first human heart transplant operation in 1967. Although Demikhov was a skilled surgeon, he did not address the problem of tissue rejection. Most of his canine test subjects died from rejection complications in the weeks after surgery; one survived up to a month.

Demikhov repeated and perfected his procedure for creating two-headed dogs over many years. This image is from an operation performed in 1967.

Stapp (1910–99) was born in Brazil to US missionaries. He studied biophysics and medicine in the United States before joining the Medical Corps of the US Army in 1944 and beginning his pioneering studies into rapid deceleration.

After World War II, aeroplanes were flying faster than ever before, which raised the question of whether the human body could endure the deceleration that a pilot might face in a crash. On 10 December 1954, US Air Force Major John Paul Stapp embarked on an experiment in the Sonic Wind I rocket sled at Holloman Air Force Base in the US state of New Mexico. The test was designed to simulate the effect of being ejected from a jet at supersonic speed. Stapp sat in an ejector seat mounted on the sled with his legs and arms strapped down. After the nine solid-fuel rockets fired, the sled accelerated to a speed of 1,017 kilometres per hour (632 mph), which set a ground speed record. When the breaking system kicked in, Stapp's body was exposed to the equivalent of forty-six times the force of gravity (46G), and when the sled stopped, Stapp reported that he could not see. Fortunately, this blindness was short lived and was caused by the bursting of the blood vessels in his eyes.

Stapp began his experiments in 1947, and rode a fast-stopping, rocket-powered sled named Gee Whiz mounted on a track. He rode this sled twenty-six times and experienced up to thirty-five times the force of gravity (35G) during the rapid deceleration.

I practised dressing and undressing with the lights out so if I was blinded [in an experiment] I wouldn't be helpless.

Stapp proved that it was possible for pilots to survive the greater forces that they would be exposed to if they ejected from aeroplanes flying at supersonic speeds. Although he had temporarily blacked out during the experiment, his record-breaking feat earned him the moniker 'the fastest man on Earth' from *Time* magazine, and he received other media attention, too. He used this to help make civilian transport less dangerous. He campaigned to make the kind of safety measures that were standard in aeroplane design available in cars. Working tirelessly, he organized conferences with manufacturers and the government, and eventually succeeded in persuading the industry to redesign seat belts to be more effective.

On the Gee Whiz rocket sled, Stapp regularly decelerated from speeds of more than 805 kilometres per hour (500 mph) to zero in as little as 1.6 seconds.

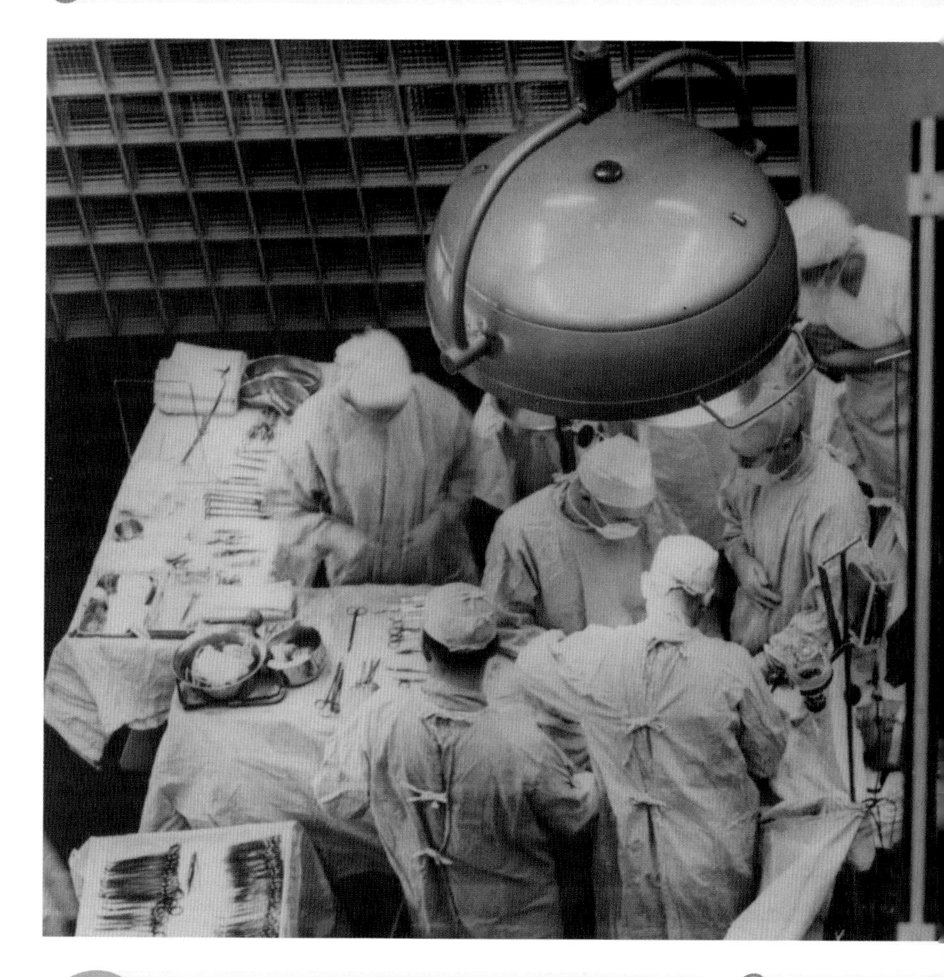

Before the mid 1950s, attempts at organ transplants had ended with the patient's body rejecting the donated tissue and their immune system attacking it as a foreign biological threat. By contrast, Richard, who did not face any rejection complications, lived eight years after the operation and had a family in that time. The success of his operation proved that organ transplants could not only save lives, but also improve the quality of life. Subsequent advances in immunosuppressant drugs enabled Murray to perform a successful kidney transplant from an unrelated donor into patient Mel Doucette in 1962.

In the surgical theatre at Peter Bent Brigham Hospital, Murray and his team worked quickly in order to maximize the chances of success.

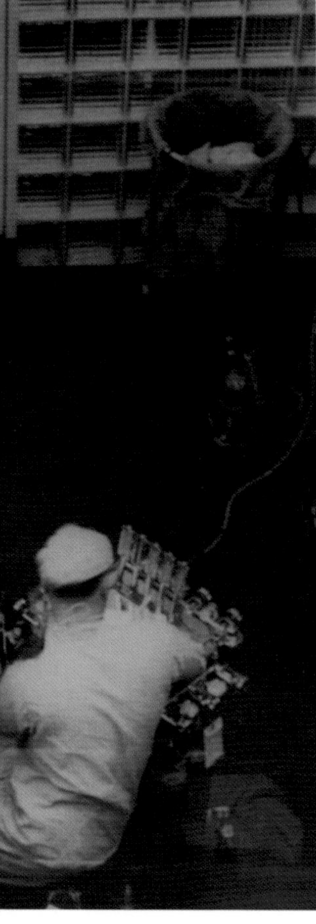

From a young age, Murray (1919–2012) wanted to be a surgeon and he attended Harvard Medical School before serving in the Medical Corps of the US Army during World War II. In 1990 he shared the Nobel Prize in Physiology or Medicine with E. Donnall Thomas for their pioneering work in the field of 'organ and cell transplantation in the treatment of human disease.'

KIDNEY TRANSPLANT

JOSEPH MURRAY

1954

In 1954, while he was working at Peter Bent Brigham Hospital in Boston, Massachusetts, surgeon Joseph Murray was referred the case of a young patient named Richard Herrick, who had been diagnosed with chronic kidney disease. Richard had a twin brother, named Ronald, who was willing to donate a kidney. In order to test the twins' genetic compatibility, Murray grafted a piece of skin from Ronald onto Richard. When this showed no sign of rejection, Murray was reassured that a successful organ transplant was possible. However, he was still nervous about getting the operation right, so requested the use of a human corpse on which to practise the procedure. Once he was satisfied that he could attach the donor kidney quickly and effectively, he scheduled the operation for 23 December 1954, and readied his team. At Murray's signal, fellow surgeon John Harrison removed the donor kidney and delivered it to Murray, who connected the organ to his patient's veins and arteries and bladder as quickly and methodically as he could. He knew that the faster he connected the donor kidney to his patient's circulatory system, the better the prognosis would be. The whole operation lasted a total of five-and-a-half hours, and the result was a resounding success.

During World War II, Murray had no permanent success when he grafted donated skin onto burns patients. However, he knew that a colleague, James Barrett Brown, had successfully grafted skin from one identical twin to the other.

Operation Whitecoat was endorsed by Harry Armstrong (1899–1983), surgeon general of the US Air Force. Today, he is best known for describing what is known as the Armstrong Limit. This is the altitude at which water boils at body temperature and humans cannot survive in an unpressurized environment.

We start off healthy . . . after recovery they continue testing to see if we come back to full strength.

VOLUNTEER

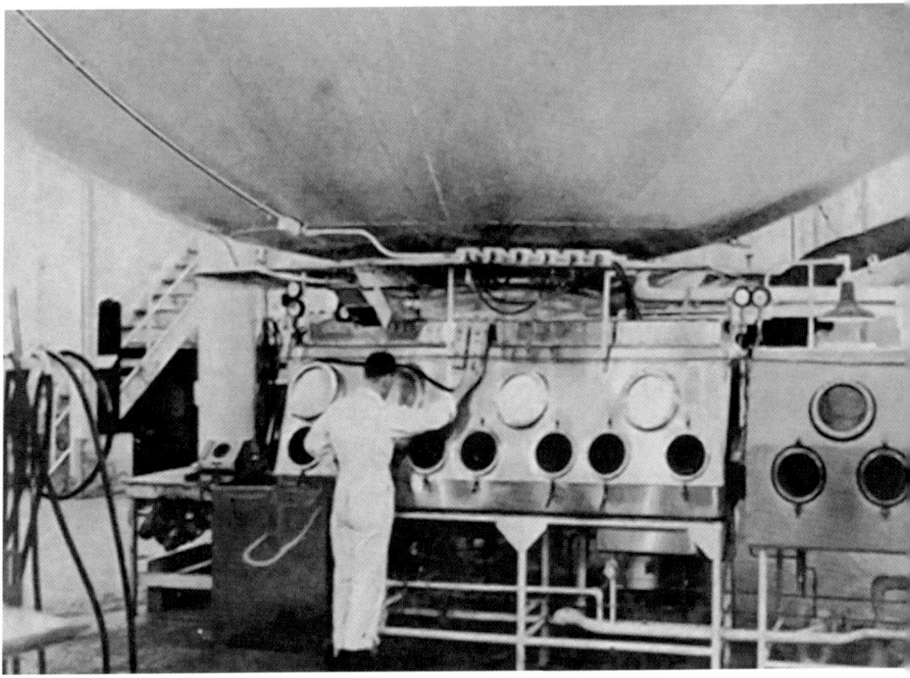

More than 2,300 volunteers participated in Operation Whitecoat tests from 1954 to 1973. A variety of pathogens was used but no fatalities were recorded during the experiments. In the course of the programme, scientists developed vaccines for West Nile fever, sandfly fever, typhoid fever and typhus. However, although the participants were told that the programme focused on defensive measures, the data it generated could also have been used in the development of offensive chemical weapons. Today, many of the surviving veterans have mixed attitudes towards their participation in the experiments, but most are proud that they were able to serve their country.

Inside the cubicle under the eight ball (the huge metal sphere at the top of the picture), Operation Whitecoat volunteers would breathe in the pathogens that were being tested for that day.

OPERATION WHITECOAT

US ARMY

1954–73

A unique deal was struck in 1954 between the military and senior figures in the Seventh-day Adventist Church. Freshly drafted conscientious objectors were invited to volunteer for Operation Whitecoat, a programme to test biological weapons on human subjects in order to develop better vaccines. The recruits were free to withdraw from the programme at any time and to become army medics instead. They were fully briefed on the details and risks of each experiment and chose which, if any, they wanted to participate in. They then signed consent forms, thus demonstrating their willingness. Some tests involved entering a cubicle and inhaling through a face mask the contents of a giant metallic sphere dubbed the 'eight ball'. This could be anything from hazardous pathogens to ordinary air. The largest scale open-air Operation Whitecoat experiment occurred on the evening of 12 July 1955, at Dugway Proving Ground in the US state of Utah. About thirty volunteers lined up, alongside guinea pigs and monkeys in cages, in front of canisters that contained Q fever pathogens. At the appointed time, the canisters sprayed their contents towards the volunteers, forming a toxic mist. Some of the volunteers had been vaccinated against Q fever and others had not, but all were instructed to breathe normally. None of the vaccinated volunteers became ill, whereas those who had not been inoculated developed Q fever within days.

Because Seventh-day Adventists typically abstain from both smoking and drinking, the volunteers were in better health than the average US Army recruit.

Gathering data on the effects of biological weapons on humans is problematic. Troops need to be prepared for the risks they may face in battle, but exposing people to dangerous substances without their informed consent is unethical. The health of the volunteers in Operation Whitecoat was monitored closely and prompt medical aid was provided to anyone who developed symptoms.

Delgado (1915–2011) joined the department of physiology at Yale University in 1946. While there, his research focused on therapeutic brain stimulation experiments. After a period of success, his work on the control of the mind through brain stimulation fell rapidly out of favour, prompting him to return to his native Spain in 1974.

STIMOCEIVER
JOSÉ MANUEL RODRÍGUEZ DELGADO
1963

From the mid 1950s, Spanish physiologist José Manuel Rodríguez Delgado began to investigate the effects of direct electrical stimulation on the brain. As part of his research, he invented a small, radio-controlled, electro-stimulation device, which he termed a 'stimoceiver', and implanted it in the brains of his test subjects. The device monitored brainwave activity and could deliver electrical stimuli directly to specific parts of the brain. In 1963 Delgado began to work with bulls that had been bred for bullfights, and he selected these animals specifically because they were known for their aggressive behaviour. He observed several bulls on a ranch in Córdoba, Spain, and first mapped how various areas of their brains responded to electrostimulation. The results were broadly in line with his previous experiments. In order to demonstrate his stimoceiver, Delgado used a bull that had the device implanted in its caudate nucleus, the area responsible for voluntary movement. He got into the ring with the bull, and when the animal ran at him, Delgado pressed the button on the device. The bull immediately halted its charge and turned to one side.

What would happen if you could physically control someone's mind? This was the philosophical question that drove much of Delgado's research. He firmly believed that his stimoceiver might offer an alternative to invasive brain surgery and, despite the serious ethical concerns of some of his critics, that it might have applications in wider society, too.

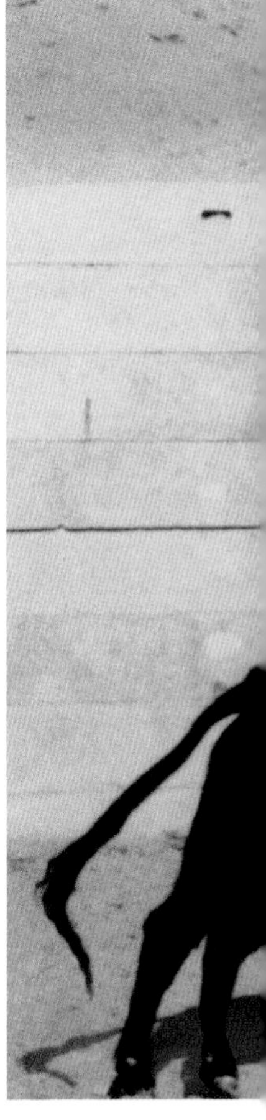

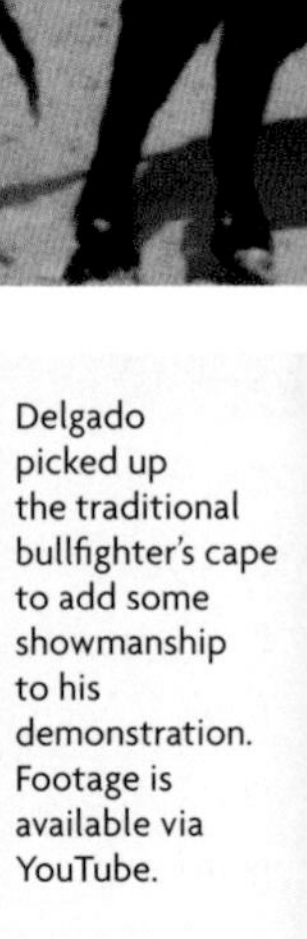

Delgado picked up the traditional bullfighter's cape to add some showmanship to his demonstration. Footage is available via YouTube.

“

The purpose is physical control of the mind. Everyone who deviates from the given norm can be surgically mutilated.

?

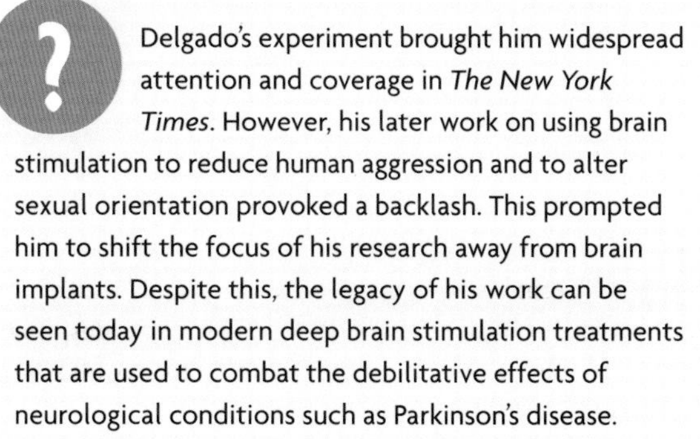

Delgado's experiment brought him widespread attention and coverage in *The New York Times*. However, his later work on using brain stimulation to reduce human aggression and to alter sexual orientation provoked a backlash. This prompted him to shift the focus of his research away from brain implants. Despite this, the legacy of his work can be seen today in modern deep brain stimulation treatments that are used to combat the debilitative effects of neurological conditions such as Parkinson's disease.

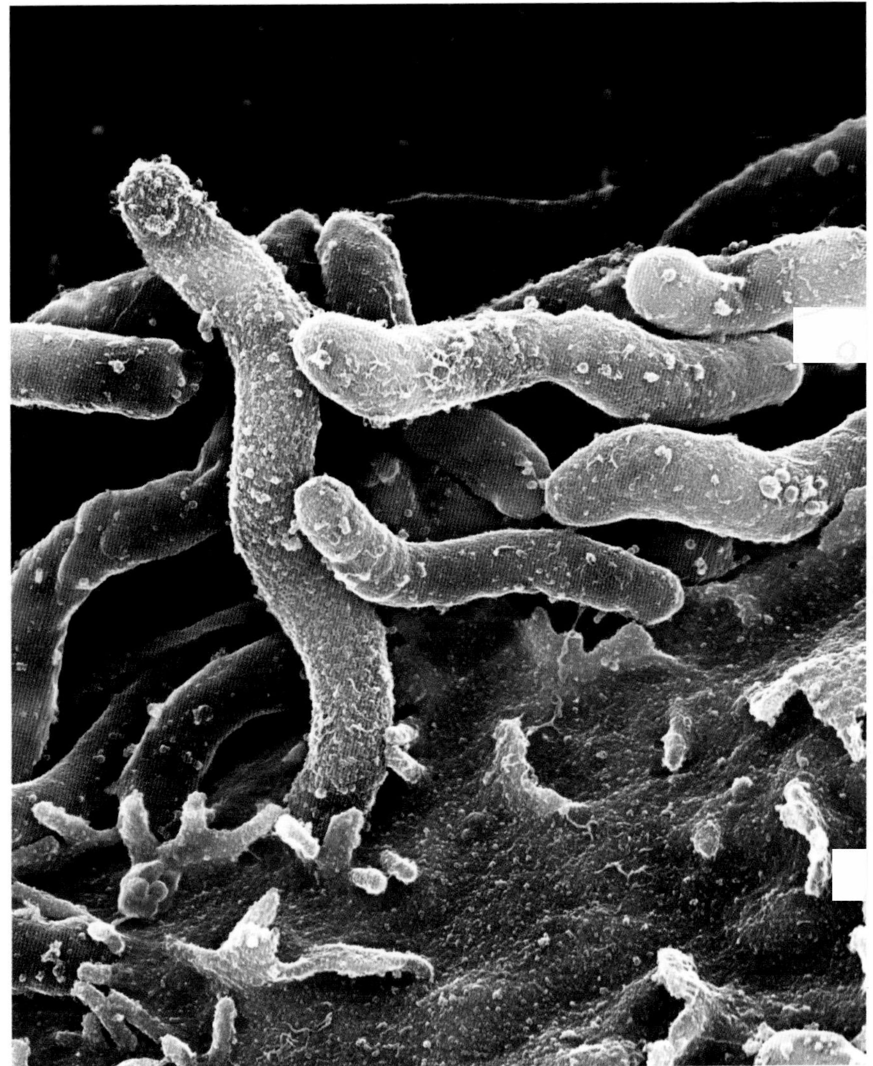

Despite the fact that Marshall's experiment established *H. pylori* as the underlying cause of stomach ulcers—fully treatable therefore with a readily available course of antibiotics—it took another decade for these results to become widely accepted. Marshall went on to develop a successful non-invasive diagnostic test for *H. pylori* infection.

In the second half of the 20th century, mainstream medics were convinced that stomach ulcers were caused by stress, and gastric surgery was a common treatment. Physician Barry Marshall, on the other hand, had observed *Helicobacter pylori* bacteria in the biopsies of several stomach ulcer patients and had effectively treated them with a combination of antibiotics and bismuth salts. He had also read medical literature and noted references that appeared to back up a link between *H. pylori* and stomach ulcers, but in animal trials his attempts to induce ulcers using the bacteria had failed (*H. pylori* affects only primates). Without the backing for human trials, Marshall opted to experiment on himself. First, he submitted to an endoscopy in order to prove that he did not already have *H. pylori* in his digestive tract. He then drank a broth containing a sample of the bacteria harvested from a patient, thereby deliberately infecting himself. After five days, Marshall felt nauseous, vomiting in the morning, and after ten days an endoscopy revealed the presence of *H. pylori* in his digestive tract, along with inflammation and all the signs of gastritis, the precursor to an ulcer. He prescribed himself his own simple antibiotics and bismuth salts treatment, which dealt with the bacteria and cured the infection.

In the late 1970s, stomach ulcers were routinely treated with antacids, which addressed the painful symptoms of stomach acid acting on the ulcers but did nothing to prevent the ulcers from recurring. Consequently, some patients lived in pain and others endured unnecessary gastric surgeries, including having their stomachs removed.

Marshall (1951–) was born in Western Australia. After graduation, he specialized in gastroenterology and met pathologist Robin Warren, who introduced him to *H. pylori*, in 1981. Together they won the Nobel Prize in Physiology or Medicine in 2005 for 'their discovery of the bacterium *H. pylori* and its role in gastritis and peptic ulcer disease'. Today, Marshall leads the *H. pylori* Research Laboratory at the University of Western Australia.

I didn't discuss it with [my wife] . . . This was one of those occasions when it would be easier to get forgiveness than permission.

In this magnified false-colour image, the *Helicobacter pylori* bacteria (coloured red) is shown in its favourite home environment: a human stomach.

The birth of Dolly the cloned sheep was a watershed moment in genetic engineering, opening up many new possibilities and seizing the popular imagination. It demonstrated that differentiated adult cells of complex creatures, such as sheep, could be reprogrammed. However, although the techniques used to create Dolly have been employed to clone other animals, and they have been adapted to create transgenic animals for use in 'pharming', scientists are still exploring avenues for their potential application with human genes.

Dolly is named after US country music singer Dolly Parton. Sheep were used for the experiment because they were cheaper and more easily available than cattle.

ANIMAL CLONING
KEITH CAMPBELL AND IAN WILMUT
1995–96

In 1995 British geneticist Keith Campbell had already used a technique called nuclear transfer to create Megan and Morag, sheep cloned from a sheep embryo. The process involved placing cells—which contain a complete set of genetic information—from a sheep embryo into an oocyte (egg cell) that had its nucleus, and therefore its own genetic information, removed. However, contrary to prevailing opinion, Campbell was sure that somatic (adult) cells could also be used to create clones, especially if he could induce them into quiescence: pausing their cycle of cell division by starving them of nutrients. Working with his senior colleague Ian Wilmut, Campbell harvested somatic cells from the udders of a white-faced Finnish Dorset sheep, induced quiescence and transferred a nucleus from these cells to the oocyte of a Scottish Blackface ewe. Campbell then used a small pulse of electricity to fuse the new nuclei with the oocyte, prompting the new cell to begin division. This process was repeated 277 times. After six days, only 29 of the 277 fused eggs had developed into embryos. These were placed into the wombs of thirteen Scottish Blackface ewes to gestate. Dolly, born on 5 July 1996, was the only clone sheep to survive birth.

After studying cell growth cycles, Campbell believed that adult cell nuclei could be implanted into a donated egg that had its own nucleus removed and result in the birth of a clone whose genes would match the adult cell.

Wilmut (1944–) has credited Campbell (1954–2012) with the greater input into Dolly's creation. In 2008 the pair won the Shaw Prize, which they shared with Japanese geneticist Shinya Yamanaka, for their 'work on the cell differentiation in mammals'. They also worked on the creation of two transgenic cloned sheep, dubbed Molly and Polly, that had the human gene for the blood-clotting protein, factor IX.

CAENORHABDITIS ELEGANS GENOME

RICHARD DURBIN AND JEAN THIERRY-MIEG

1998

The complete set of genetic instructions, or genome, for the roundworm *Caenorhabditis elegans* consists of about one hundred million base pairs (far fewer than the human genome). Sequencing that genome would necessitate a collaborative effort involving teams of scientists, each cutting, copying and dyeing small sections from the genome as they attempt to identify, map and sequence each nucleotide. Consequently, such an experiment required the development of a new tool to distribute and coordinate the vast amount of data involved. Together, geneticists Richard Durbin and Jean Thierry-Mieg developed a custom-made and adaptable database for the project, with a Windows-based graphical user interface. Dubbed AceDB (a *C. elegans* database), the database was a significant step in the evolving field of bioinformatics, and facilitated the efficient collation and dissemination of genetic data between researchers. The program also included tools for researchers to annotate and make calculations concerning various elements of their findings. The proof of the success of AceDB came in 1998, when *C. elegans* became the first animal to have its genome fully sequenced for publication.

As an informatics specialist, Durbin (1960–) greatly increased the efficiency of the *C. elegans* genome project. His background in mathematics and his doctoral thesis on the development and organization of the nervous system of *C. elegans* roundworm left him ideally placed to develop the appropriate tools to help researchers share data. In addition to his work on *C. elegans*, Durbin received the Royal Society's Mullard Award in 1994 for his contributions to the improvement of laser-scanning confocal microscopy. He was elected as a fellow of the Royal Society a decade later, in 2004, in recognition of his work in bioinformatics.

In 1963 South African biologist Sydney Brenner first proposed *C. elegans* roundworm as an ideal model life form in a variety of science experiments because it is a small, simple, transparent creature that is easy to work with. Between 2014 and 2015, *C. elegans* on board the International Space Station were used to study the epigenetic expressions of four generations bred in space.

The transparent roundworm *C. elegans* can reproduce, either sexually or asexually, after only four days. It has an average lifespan in the laboratory of two to three weeks, which makes it ideal for research.

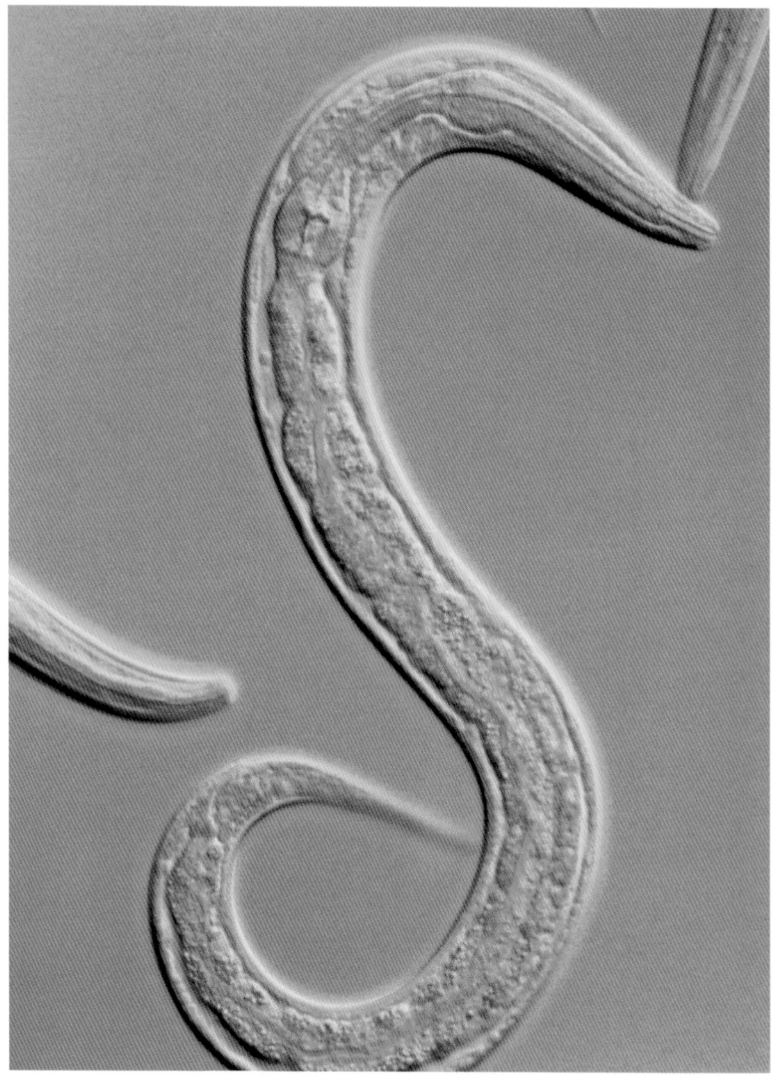

The *C. elegans* genome project provided researchers with a map of the creature's genome, which could be used to guide future genetic experiments using the roundworm. The charting of the worm's cell development has provided insights into the process of programmed cell death, which has implications for human diseases, including some cancers, that involve the failure of this mechanism.

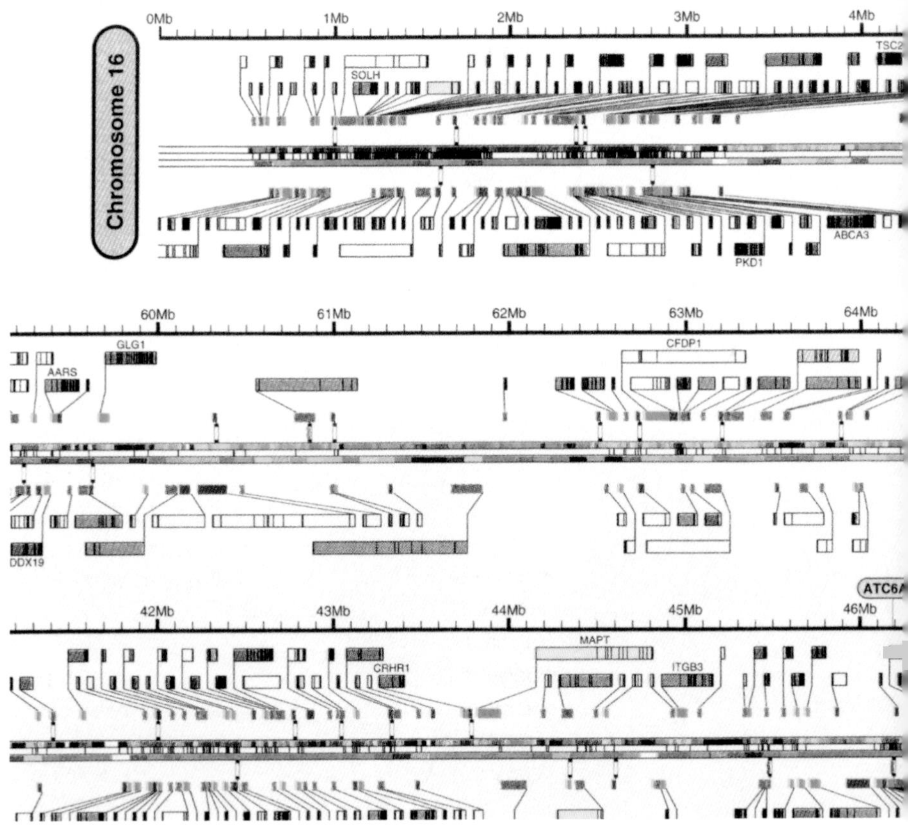

This project has developed different techniques and facilities for mapping and sequencing genomes for humans, which has, in turn, speeded up the process of sequencing genomes for a variety of other organisms. It has also provided researchers with the sequence of almost all the chemical base pairs that form the genome, as well as the physical and genetic maps of its structure. This data helps scientists to better focus their research, to isolate the genetic factors related to various conditions and even to suggest new treatments and therapies. It also reveals that we still have much to learn about our genes, including the reason why humans appear to have far fewer active protein-coding genes than was previously thought.

Chromosome 16, pictured here, contains about 90 million base pairs and represents just less than 3 per cent of the 3.2 billion base pairs in the human genome.

Venter (1946–) is best known as one of the first scientists to map the human genome. He continues to work on a variety of gene-related projects, from biofuels to developing synthetic organisms. He also runs the J. Craig Venter Institute, founded in 2006.

HUMAN GENOME PROJECT
CRAIG VENTER
2003

Mapping the human genome, the complete set of instructions on how to build a human—both in terms of where the genes appear and what they do—required teams of scientists to work together to sequence all 3.2 billion base pairs (either guanine-cytosine/GC or adenine-thymine/AT) that form each rung on the ladder of human DNA. Using techniques pioneered by British biochemist Frederick Sanger, project scientists laboriously cut, cloned and dyed short sections of DNA to sequences of fewer than 1,000 base pairs. Slowly and systematically they began to work up the whole DNA sequence. US geneticist Craig Venter was originally part of the publicly funded human genome project, but became frustrated with the slow progress and left to lead the private company Celera Genomics. He wanted to identify genes more quickly and more effectively, and eventually to patent them. Together with his team of scientists, he employed a different technique, termed 'shotgun sequencing', which involved shredding multiple copies of the target DNA at random into small pieces. These pieces were then sequenced and fed into a computer to reconstruct the overall base pair sequence (and identify any gaps) based on areas of overlap in the multiple shredded copies. The first drafts of the human genome were published by both the public and private projects in 2001, and the full sequence in 2003.

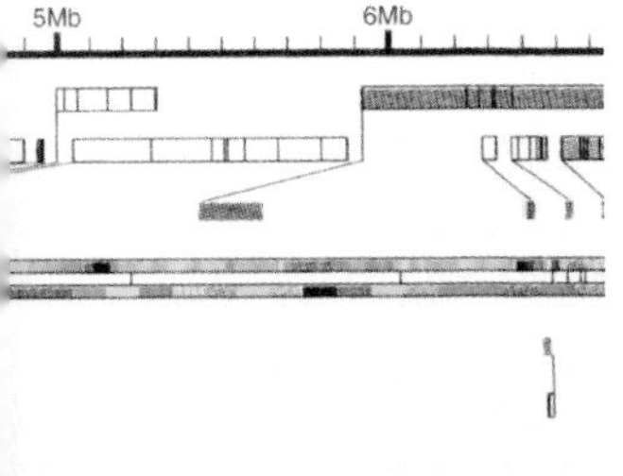

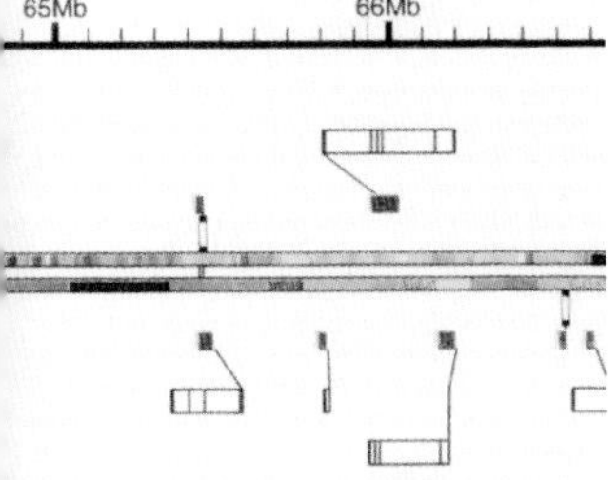

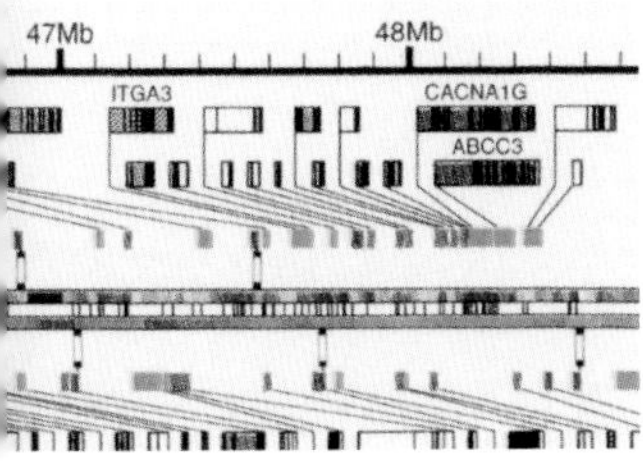

The systematic approach to mapping the human genome began in the 1980s at the US Department of Energy, which sought to improve its ability to detect genetic mutation caused by radiation. A fully mapped 'normal' human genome would help to define and identify abnormal mutations.

ON
OFF
MAIN POWER

VOLTAGE
ENERGIZER

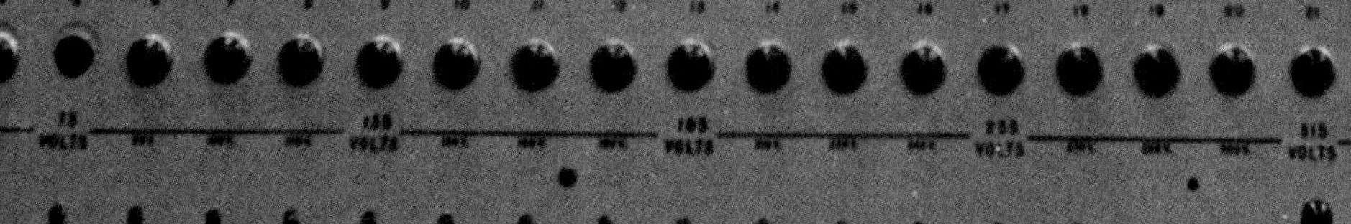
75
VOLTS
135
VOLTS
195
VOLTS
255
VOLTS
315
VOLTS

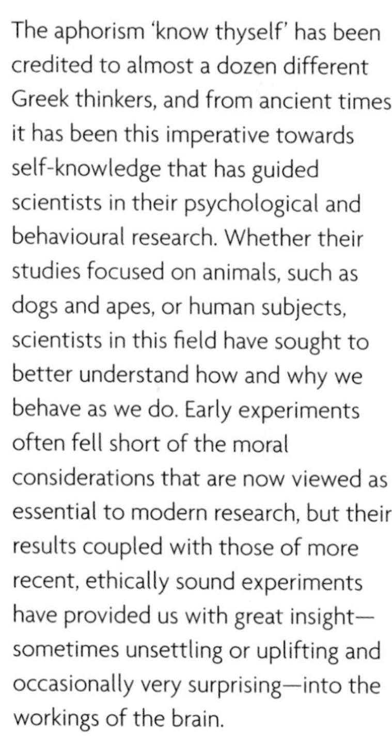

CHAPTER TWO
PSYCHOLOGY AND BEHAVIOUR

The aphorism 'know thyself' has been credited to almost a dozen different Greek thinkers, and from ancient times it has been this imperative towards self-knowledge that has guided scientists in their psychological and behavioural research. Whether their studies focused on animals, such as dogs and apes, or human subjects, scientists in this field have sought to better understand how and why we behave as we do. Early experiments often fell short of the moral considerations that are now viewed as essential to modern research, but their results coupled with those of more recent, ethically sound experiments have provided us with great insight—sometimes unsettling or uplifting and occasionally very surprising—into the workings of the brain.

‹ Obedience to Authority (see p. 76)

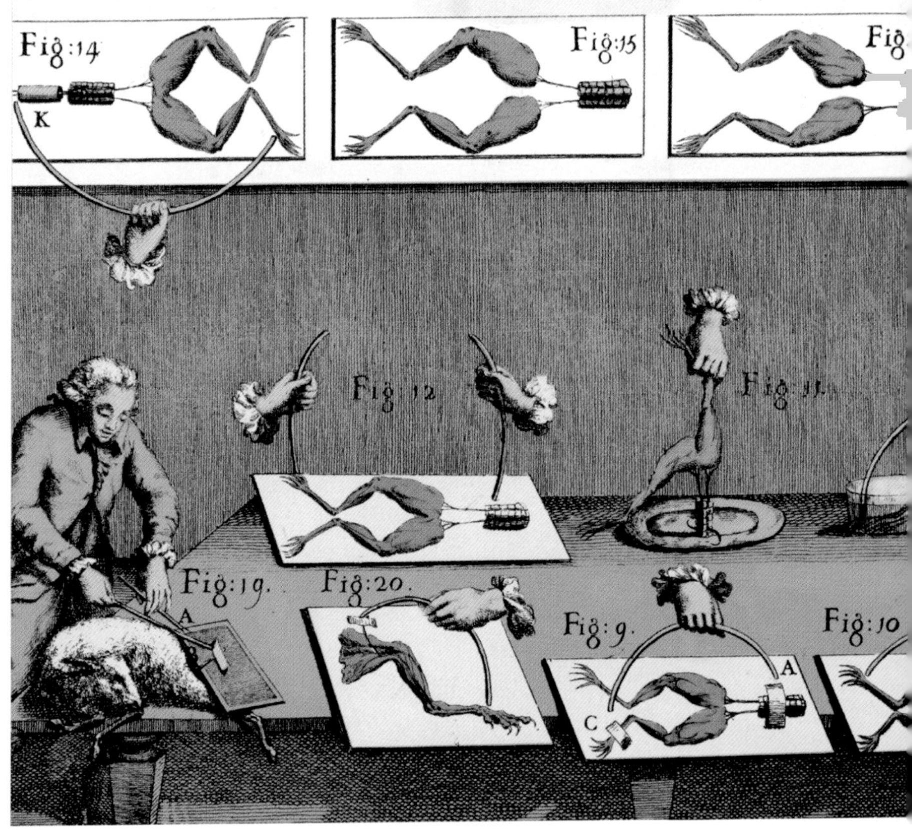

Galvani's discovery greatly excited scientists of the day, including Alessandro Volta, who initially confirmed his peer's results. However, Volta remained sceptical as to whether animal electricity was intrinsic to the muscles of animals. He postulated that the source was in fact a reaction between the animal's body fluids and two types of metal. He therefore defined animal electricity as a product of what he termed 'Galvanism': an electric current created by a chemical action. Although the two scientists disagreed, they were respectful of one another's work. Today, Galvani is credited with the discovery of bioelectricity, and its existence underpins the workings of the central nervous system and how it is controlled by signals from the brain.

Inspired by his own observations, Galvani conducted a variety of experiments applying static electricity to the anatomy of dissected frogs.

In the 18th century, electricity was a phenomenon that fascinated scientists. They studied its properties and 'stored' it in devices such as the Leyden jar, but Galvani was the first to suggest the existence of bioelectricity (electricity produced by a living organism).

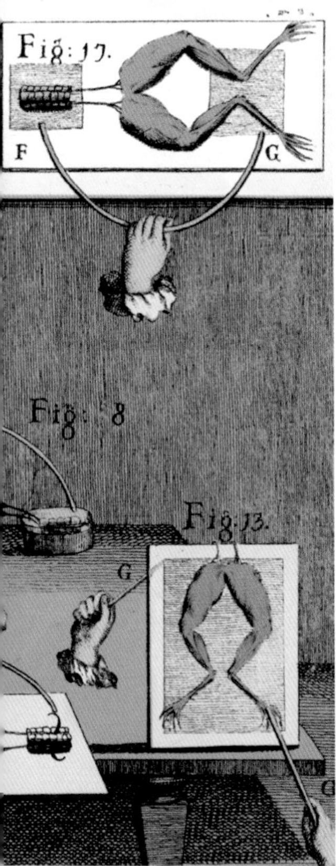

ANIMAL ELECTRICITY
LUIGI GALVANI
1780

According to his own accounts, Italian physician and biologist Luigi Galvani discovered bioelectricity quite by accident. One day in 1780 he was in his laboratory preparing to work on a dissected frog when his assistant touched a nerve in the animal's leg with his scalpel and caused it to twitch. Galvani repeated his assistant's actions and also induced a physical response. He speculated that the metal hooks used to hold the frog in place on the iron lattice were drawing atmospheric electricity into the frog's body. However, when he left dissected frogs outside to see if such electricity would cause the same twitching, the results were inconclusive. Back indoors he fixed the frogs to iron plates using brass hooks through their 'spinal marrow', and again probed the frog with his scalpel, which caused the animals to convulse. Galvani concluded that the movement was caused by some kind of 'fluid from the nerves' conducted through the metal. He subsequently referred to this fluid as animal electricity.

It is easy in experimenting to deceive ourselves, and to imagine we see the things we wish to see.

Galvani (1737–98) worked at the Academy of Sciences of the Institute of Bologna. His experiments with animal electricity inspired Mary Shelley's *Frankenstein* (1818) and his name was given to the galvanometer, a device for detecting electrical currents.

The science of psychology was just beginning to establish itself at the dawn of the 20th century. However, Pavlov's experiment, which became one of the field's most iconic and recognizable, was not designed to test a psychological theory, but arose as a consequence of studies into the process of digestion.

CONDITIONED REFLEX
IVAN PAVLOV
1902

As professor of physiology at the Imperial Medical Academy in St Petersburg, Ivan Pavlov began to explore the digestive systems of dogs in 1890. Unlike many of his contemporaries, he was committed to studying live rather than vivisected subjects, and devised a surgery to perform on the animals that made it easier for him to access their stomachs for research. As part of his work, Pavlov recorded which stimuli would prompt a hungry dog to salivate: a behaviour that is an unconditioned part of the digestive process. He noticed that the animals would often salivate as soon as the assistant who fed them arrived, in other words before the food was actually presented. Intrigued by this phenomenon, which he termed 'psychic secretion', Pavlov decided to test the dogs' responses to see if their behaviour could be conditioned. He introduced a procedure of ringing a bell before the animals received their food. The dogs learned to associate the bell with the food that followed, and salivated upon hearing the chime in anticipation of being fed. Eventually, Pavlov tried ringing the bell without producing any food and noticed that the dogs still salivated. He repeated various versions of this experiment—using light, touch and pitch—in order to ascertain whether the dogs could learn, or be conditioned, to exhibit new behaviour responses to different stimuli.

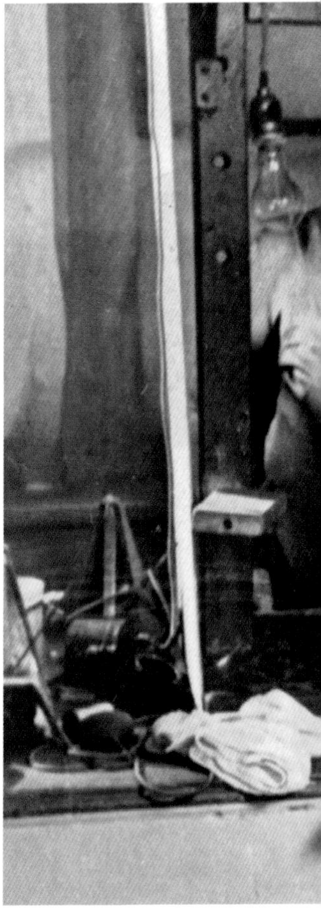

A very skilful surgeon, Pavlov (1849–1936) was well-respected during his lifetime for his physiological studies. He won the Nobel Prize in Physiology or Medicine in 1904 for his work on the physiology of digestive secretions.

Pavlov discovered that dogs could be conditioned to associate a range of stimuli with certain outcomes. This is a vital survival skill because the ability to connect warning signs to unpleasant experiences helps to avoid potential pain and danger. For humans, the most prominent use of Pavlov's conditioning theory is in aversion therapy, in which people are taught to associate discomfort with a stimulus in order to change their future behaviour towards it. It also has applications in drug and substance abuse rehabilitation programmes in which participants are attempting to modify their previous behaviours.

Pavlov remained active in his field right up until his death in 1936. Here, he is seated to observe a dog experiment in 1934.

Watson was interested in conditioned behaviour and was influenced by Ivan Pavlov's experiments on classical conditioning. He believed that, at birth, humans have a very limited range of unlearned emotional responses and sought to test if it was possible to condition fear responses in a young child.

CONDITIONED FEAR
JOHN WATSON
c. 1920

US psychologist John Watson's seminal study on conditioned fear is best known as the Little Albert experiment after its sole test subject, identified by Watson as 'Albert B . . . the son of a wet nurse in the Harriet Lane Home for Invalid Children'. When Albert was only nine months old, Watson and his assistant, Rosalie Rayner, exposed him to several animals, including a white rat, rabbit and monkey as well as a variety of masks that he had not encountered before. None of these caused the infant any concern at this stage. Watson then began to strike a steel bar with a hammer just behind Albert until he burst into tears. When the baby was eleven months old, Watson showed him the rat again, and this time he hit the steel bar whenever Albert motioned to touch the animal. This startled the infant and he recoiled, but he did not cry. A week later the test was repeated and this time he was reluctant to reach for the creature. When the rat was brought closer to Albert, Watson struck the bar. The process was repeated several times and eventually the baby burst into tears. Subsequently, he started to cry and show fear as soon as he saw the rat, without the stimulus of the loud noise. In later experiments Albert began to display what was described as a 'negative reaction' to other animals and objects to which previously he had shown no aversion.

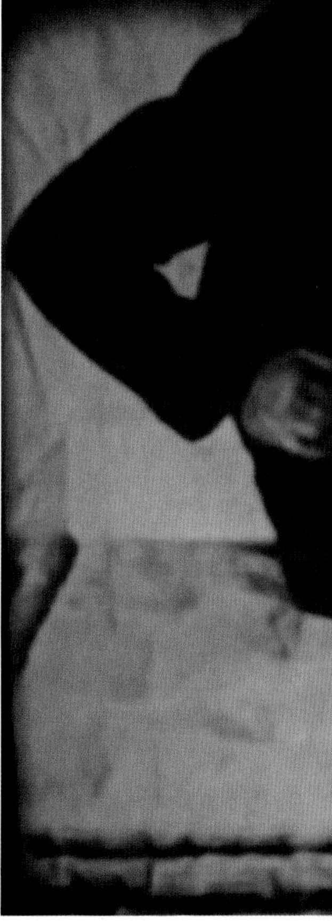

Watson (1878–1958) is seen as the founder of behaviourism, a branch of psychology that is focused on observable behaviour. His own behaviour led to him losing his position at John Hopkins University, when the affair he was having with his assistant became public.

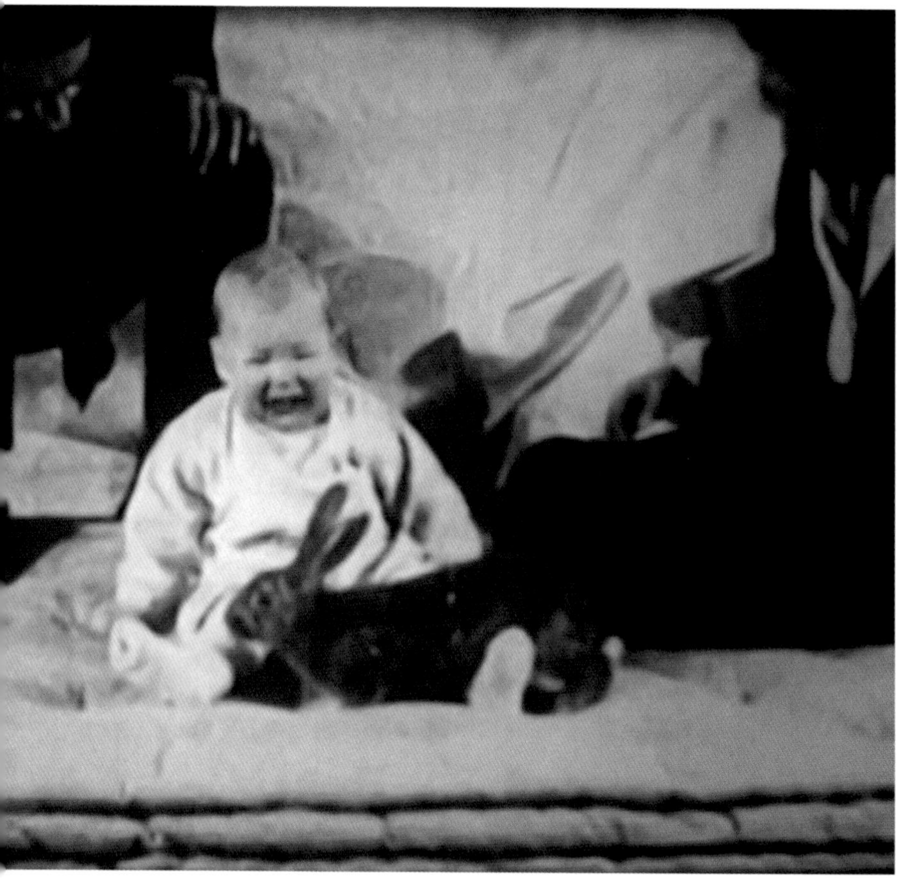

Watson's experiment is considered controversial, not only because of the small sample size and the cruelty involved in conducting experiments on a baby, but also because it ended prematurely, before Watson and Rayner could remove the conditioned fear they had instilled in Albert. However, it demonstrated for the first time that the conditioned responses that Pavlov had discovered in dogs could be replicated in humans, and it remains a classic experiment that is referenced in psychology teaching today.

We felt that we could do him relatively little harm by carrying out such experiments.

This still from the footage of the experiment shows Albert's negative reaction to a rabbit.

The purpose of the ape and child experiment was to ascertain to what extent environment played a role in primate development and socialization. Although this highly ambitious study of nature versus nurture, the first of its kind, did not focus exclusively on language acquisition and production, it greatly influenced subsequent studies in this field. In addition, it established the importance of physiological differences between humans and non-human primates in the development of verbal language.

Although pictured here holding hands, Donald the child and Gua the chimpanzee were usually tested separately during the experiment.

NATURE VERSUS NURTURE

WINTHROP KELLOGG

1931–32

Thought to be inspired by tales of the 'wolf children of India'—two girls raised by wolves who adopted wolf-like behaviour—US psychologist Winthrop Kellogg devised an experiment with his wife, Luella, in which they aised a baby chimpanzee, Gua, alongside their own baby on, Donald. The two infants were brought up as brother ınd sister and treated equally. The experiment started n June 1931 and was conducted at the family home. For welve hours a day, seven days a week, Kellogg carefully :atalogued both Gua and Donald's development, observing heir reflexes, locomotion, problem solving, play ›ehaviour, climbing, obedience, language comprehension ınd vocalization. At the age of one, Gua was able to espond more effectively than Donald to a set of twenty imple commands, and she could climb and jump better han her human sibling. Gua assimilated the colours of lothing and smells more readily, whereas Donald was ›etter at recognizing human faces. Neither infant learned o talk during the experiment. Gua vocalized her desire or food in calls that Donald understood and imitated, ›ut she was unable to form words. It is speculated that)onald's lack of progress in speaking contributed to the ›xperiment's abrupt end after nine months.

Before the Kelloggs' experiment, teaching primates human speech had focused on adult subjects. In the late 1940s and 1950s, psychologists Keith and Cathy Hayes raised a very young chimpanzee in their home and taught her to vocalize four words. However, they were only able to do this initially by moving the animal's lips for her. Later studies showed more success in teaching apes to use sign language.

Kellogg (1898–1972) was born in Mount Vernon, New York. He gained his doctorate in psychology in 1929 and soon after joined the staff at Indiana University, where he focused on conditioning and learning, mainly in animals. His ape and child experiment brought him a level of recognition that overshadowed much of his later work, including his studies on conditioning and learning in dogs and his research on bottlenose dolphin navigation, which investigated their use of echolocation.

The Kelloggs published their findings and methodology in a popular book titled *The Ape and the Child* (1933).

Gua, treated as a human child, behaved like a human child except when the structure of her body and brain prevented her.

TIME MAGAZINE

NEURAL STIMULATION

WILDER PENFIELD

c. 1937

By 1937, neurosurgeon Wilder Penfield had devised a surgical procedure for mapping the cerebral cortices of epileptic patients. His technique required the subject to be placed under local anaesthetic so that they remained conscious throughout the operation. In this way, they were able to describe their perception of their own motor and sensory reflexes in response to Penfield's electrical stimulation. The patient's head was shaved and their scalp locally anaesthetized, before the skin was peeled back and the skull cut open to expose the brain. Penfield then used unipolar or bipolar platinum electrodes, encased in plastic handles, to electrically stimulate the region of the brain around the central fold that separates the parietal and frontal lobes: the area that controls bodily movement and receives sensory input. Working methodically, he probed areas around the fold and talked to the patient at the same time. Some stimulations provoked an involuntary movement or sudden sensation, or even the early signs of an epileptic seizure. Penfield noted the precise nature of each response and later returned to probe the same area again to ensure that his stimulations produced consistent results.

Although he was born in the United States, Penfield (1891–1976) was described during his lifetime as the greatest living Canadian for his work in the field of neurosurgery. He was recruited in New York to found the Montreal Neurological Institute and Hospital at McGill University, and later renounced his US citizenship to become a naturalized Canadian. The surgeon also gave his name to the double-ended Penfield dissector, a neurosurgical instrument that is still used today. In 1960 he was awarded a Lister Medal from the Royal College of Surgeons of England 'in recognition of distinguished contributions to surgical science'.

Brain surgery is a terrible profession. If I did not feel it will become different in my lifetime, I should hate it.

In the early 20th century, the function of the brain was poorly understood, which made the treatment of neurological diseases—often collectively referred to as epilepsy—difficult and dangerous. One false move while operating on the brain could leave the patient paralysed or unable to breathe. The most common approach to such conditions was to prescribe sedatives as an alternative to invasive medical intervention, which might do more harm than good.

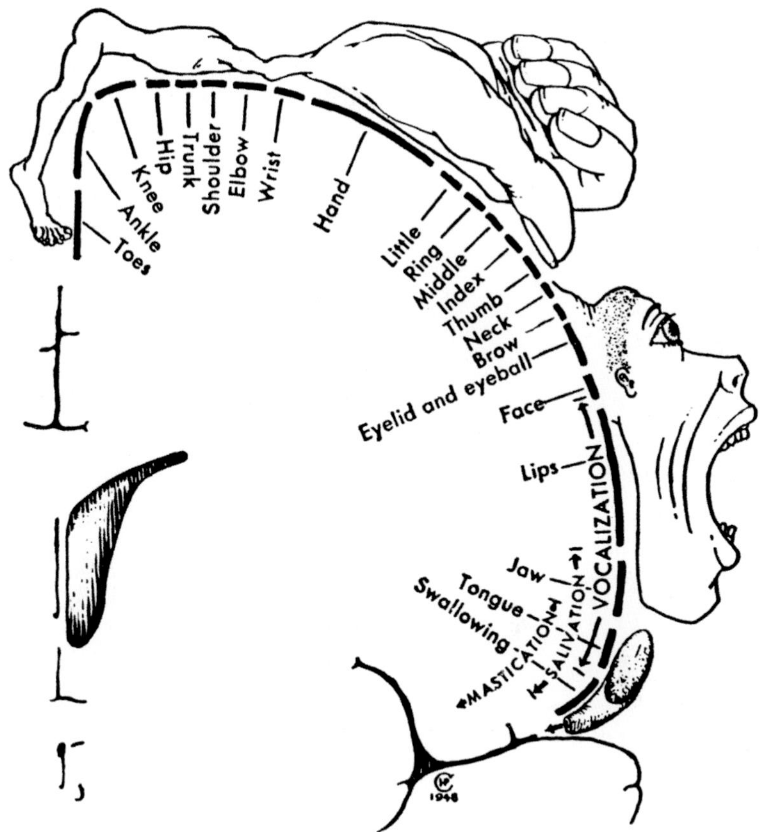

By systematically probing the brain of a conscious patient, Penfield established which areas of the organ were linked to that person's epileptic seizures. He could then produce a 'map' of the sensory and motor cortices; isolate and excise only the problematic region; and preserve the areas that were linked to other functions. This methodology, developed with his colleague Herbert Jasper, is known as the Montreal Procedure. Penfield later published generalized maps—cortical homunculi—that demonstrate which areas of the brain are linked to motor function and sensory input. They are still in use today.

This 'map' from *The cerebral cortex of man: a clinical study of localization of function* (1950) by Wilder Penfield and Theodore Rasmussen depicts the motor homunculus.

The Monster Study would not satisfy ethical requirements for psychological experiments today because of its use of child participants and the lack of informed consent. It apparently had a lasting effect on the six children in group IIA, and in 2007 they sued the State of Iowa for its negligence in allowing the experiment to be performed. The group was awarded $925,000 for lifelong psychological and emotional scars caused by their participation in the study. During the experiment, Johnson's colleagues had criticized his methods. Consequently, he did not write up his results for publication and did not reference them in his later publications. The only first-hand account of the experiment comes from Tudor's thesis.

The University of Iowa building that houses the department for speech and hearing disorders is named after Wendell Johnson.

Johnson (1906–65) survived the controversy surrounding his five-month experiment and went on to have an influential career as a psychologist. He served as director of the University of Iowa's speech clinic from 1943 to 1955.

STUTTERING IN CHILDREN

WENDELL JOHNSON AND MARY TUDOR

1939

In January 1939 graduate student Mary Tudor embarked on a study of stuttering in children; it would later become known as the 'Monster Study' for the lasting effect it had on some of its participants. The experiment was directed by University of Iowa psychologist Wendell Johnson and it involved twenty-two children from the Iowa Soldiers and Sailors Orphans' Home. Within the group, ten children were identified as having some kind of hesitancy or stutter in their speech. These ten were split into two groups: half were assured that their speech was not too bad, and the other half were told that their speech was 'as bad as people say'. The remaining twelve participants, who had no noticeable speech problems, were selected at random to complete the group. These twelve were also split into two groups: half of them were complimented on the way they spoke, and the other half—'group IIA'—were told that they had a stutter, that this was undesirable and that they should try hard to speak fluently. Tudor then informed the staff about the six children in group IIA who had been labelled as stutterers, and encouraged them to record and correct any errors these children made. At least once a month, Tudor returned to the orphanage to monitor the study. The children in group IIA were clearly the most affected. They were fractious, less communicative and more hesitant, while their schoolwork deteriorated. But, they did not stutter.

As a stutterer, Johnson was interested in whether stuttering and hesitation in speech were learned traits. He was keen to discover whether children with no previous history of such activity could be induced into stuttering and hesitation if labelled as stutterers by official figures.

Born in Warsaw in modern-day Poland, Asch (1907–96) was a leading pioneer in the field of social psychology. His experiments on prestige suggestion, in which quotations were assigned different meanings depending on the view of those to whom they were attributed, were widely lauded. The Solomon Asch Center for Study of Ethnopolitical Conflict at Bryn Mawr College in Pennsylvania is named after him, and Swarthmore College, where he was based when he performed his conformity experiments, set up an annual award in his name to recognize the most outstanding independent work in psychology at the institution.

Asch used nine different test cards, each of which showed one line on the left that matched the length of one of the three lines on the right. He showed each image to the group twice in each twenty-minute experiment.

SOCIAL CONFORMITY
SOLOMON ASCH
1951

Solomon Asch's conformity experiment was conducted on white male college students at three academic institutions, each with a differing mix of upper and lower middle-class populations. The experiment was ostensibly a test of visual acuity, and volunteers were shown cards that had four lines printed on them. They were asked to match the length of the line on the left to one of three lines on the right. In each group of up to ten volunteers, all but one participant had been instructed to give a scripted answer to the question, which they said out loud when called upon. Very occasionally these scripted answers matched the lines accurately, but most of them were quite obviously incorrect. The remaining volunteer, dubbed the 'critical subject', was unaware of the script and was seated at the end of the group. Consequently, he heard the individual answers of the other group members before he was called upon to render his own judgment. Asch hypothesized that individuals would not conform with something that was clearly wrong. He also conducted a control experiment, in which there was no pressure on the subject to conform with the rest of the group.

In the period after World War II, there was considerable interest among psychologists and the military in the extent to which group pressure affected the judgment of an individual. With funding from the US Office of Naval Research, Asch decided to develop an experiment to test the independence of an individual young man, or his public level of conformity, in the face of the unanimity of his peers whose judgment was quite clearly and factually wrong.

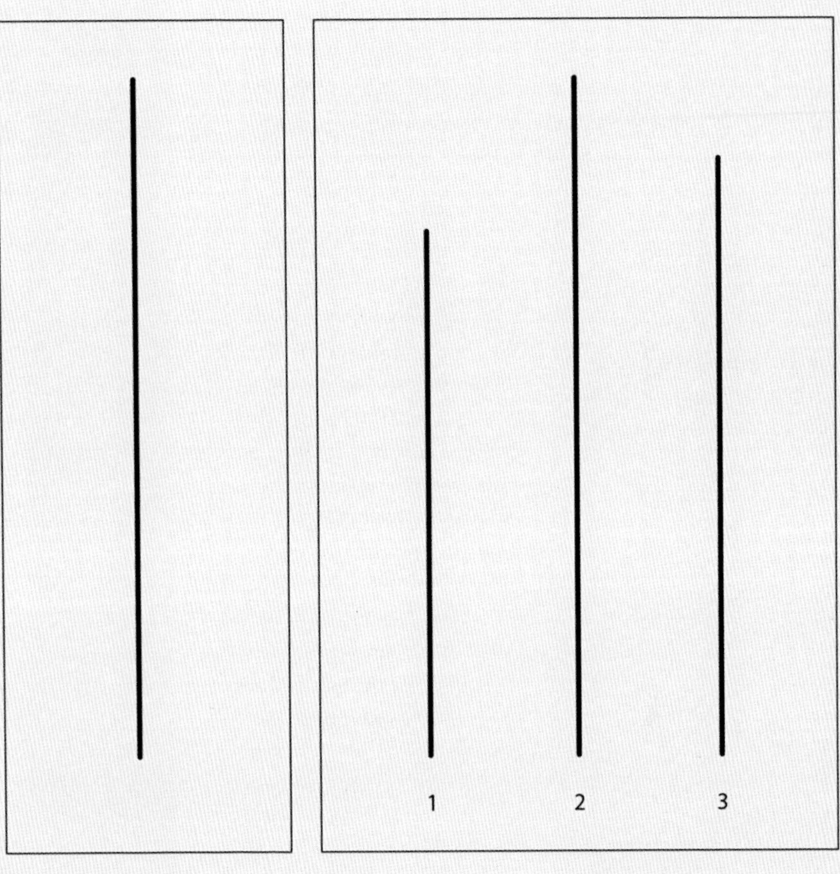

The results and post-test interviews revealed that virtually all of the 123 critical subjects noticed the inaccuracy of the answers given by their peers, but about one-third of them, a significant minority, still gave the same wrong answer in order to conform with the group. Asch's experiment proved to be a strong influence on a number of subsequent studies into conformity and obedience, particularly the work of Stanley Milgram, who was Asch's PhD student at Harvard University.

We did not consider it advisable or justified to allow subjects to leave without receiving an explanation of the procedure and of the reasons for the investigation.

Turkish American psychologist Sherif (1906–88) is a well-respected founding figure in the field of social psychology, largely due to the influence of his Robbers Cave experiment. In 1978 he was awarded the first Cooley-Mead Award from the American Sociological Association for his contributions to the field.

GROUP IDENTITY
MUZAFER SHERIF
1954

In 1954 a team of psychologists led by Muzafer Sherif organized a summer camp-like trip to Robbers Cave State Park in Oklahoma for twenty-two eleven-year-old boys. All the boys came from similar conventional family backgrounds, but they did not know each other. The excursion was, in fact, a three-phase experiment. In the first stage, the boys were gathered into two groups, each unaware of the other's existence, and taken to different areas of the park. Each group was tasked with coming up with a team name, and they chose 'The Eagles' and 'The Rattlers', respectively. In the second stage, the two groups were brought together and set in opposition, as the teams competed against one another in various contests. Throughout this part, the researchers monitored the boys' interactions and conflicts, both during the contests and afterwards, noting how the participants showed loyalty to their own group and varying levels of hostility to their opposition. In the final stage, the two groups were faced with a series of challenges that required cooperation in order to achieve a positive result. In this last phase, the antagonism and old group identities gradually broke down.

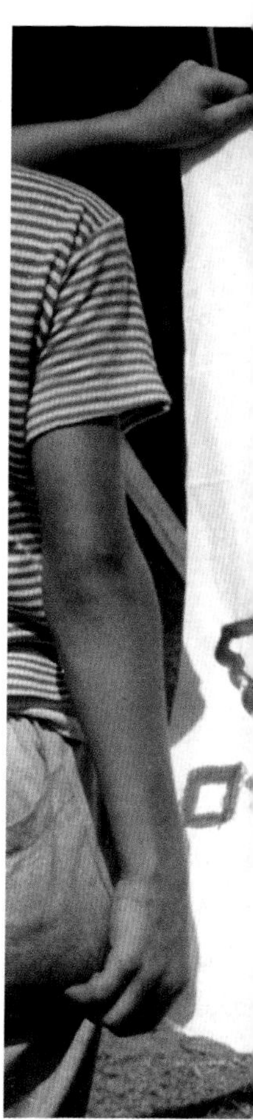

During the competition stages of the experiment, boys in each group—the Eagles and the Rattlers—designed banners to cheer on their teammates against the opposition.

The Robbers Cave experiment, as it became known, grew out of Sherif's earlier studies on the influence that groups have on perception. It was designed to see how quickly a group identity could be formed among strangers, how that identity was tested by or reinforced in competitive settings against another group, and how the group conflict could be overcome subsequently.

> *The Eagles burned The Rattlers' flag. The Rattlers ransacked The Eagles' cabin, overturned beds and stole private property.*
>
> *SOCIAL PSYCHOLOGY* (1997)

The most uplifting aspect of this experiment is that the groups found routes to overcoming prejudice and conflict on smaller scales through the adoption of wider tasks. In 1961 Sherif and his wife published *The Robbers Cave Experiment: Intergroup Conflict and Cooperation,* which developed his own observations into the framework of realistic conflict theory. The book explores how intergroup conflict arises from intergroup competition for desired objects and resources, rather than individual differences.

?

These experiments were unquestionably cruel, but Harlow maintained that the misery of a few monkeys was justified if he could convince parents to show their human children the love they needed. He used sensational terms—'love' instead of 'attachment', and 'the pit of despair'—to describe his experiments, and admitted 'the only thing I care about is whether the monkeys will turn out a property I can publish.'

MOTHER SURROGATES

HARRY HARLOW

1958

At the University of Wisconsin, US psychiatrist Harry Harlow embarked on a study to explore whether infant monkeys required more than sustenance to thrive in captivity. The experiment was designed to test if they needed comfort from a maternal figure. In a cage he placed two crude 'mother' figures: one was made of bare wire and contained a feeding apparatus; the other was constructed using the same wire, this time covered with a soft towelling fabric. It also had a friendlier face. The monkeys needed to visit the wire mother to feed, but Harlow noted that they only did so out of necessity, and in short bursts. They always returned to the comfort of the cloth mother and spent up to eighteen hours a day close to her, compared to less than an hour on the wire mother. Harlow was intrigued that a comforting mother figure trumped a source of food and so conducted follow-up experiments. In one, a resting monkey was exposed to a noisy, scary toy robot, which prompted the infant to seek refuge with its cloth mother. A further experiment involved various modified cloth mothers that emitted disturbing high-pitched noises or even spikes that repelled the infant. However, as soon as these unpleasant stimuli ceased, the young monkey would return to the security of its ersatz mother.

In 1932 Harlow established a breeding programme for rhesus monkeys to supply infants for experiments. In order to reduce the risk of them contracting diseases, he separated the young from their mothers. Despite being healthy and well fed, they became anxious. They also became attached to the soft fabric lining of their cages, an observation that prompted further study.

Despite much criticism about the ethics of his experiments, Harlow (1905–81) received considerable recognition for his work throughout his career. He won the Howard Crosby Warren Medal in 1956 for his achievements as an experimental psychologist, the National Medal of Science in 1967 and the Gold Medal from the American Psychological Association in 1973. Outside of psychology, he served as the head of human resources research at the US Department of the Army for three years in the early 1950s.

An infant rhesus monkey clings to the cloth mother Harlow designed to test a monkey's need for comfort.

I don't have any love for them. Never have. I don't really like animals. I despise cats. I hate dogs. How could you like monkeys?

The Milgram experiment was controversial at the time and would be considered unethical today because of the lack of informed consent from its participants. However, it remains a highly influential work not only for what it purports to tell us about obedience—and opinion on that remains divided—but also for its flawed design, which has been criticized for everything from its limited scope to the way in which, at the end of the experiment, it gave volunteers the impression that their actions were scientifically valuable and therefore justified.

Stanley Milgram is pictured here with the shock generator. The switches were clearly labelled to indicate how severe a shock would be administered.

OBEDIENCE TO AUTHORITY

STANLEY MILGRAM

1961

US psychologist Stanley Milgram conducted this experiment using forty male volunteers recruited through newspaper advertisements. They thought that they were participating in a study on the effect of punishment on learning. Instead, the experiment's purpose was to test obedience to authority. The men were paired with one of Milgram's confederates, each of whom presented himself as another volunteer but, in fact, provided only scripted responses throughout the experiment. The true volunteers were told that they would perform the role of the teacher and they were placed in a separate room to the confederate, or learner. This room was equipped with a 'shock generator' that had a variable output dial labelled from '15V—slight shock' up to '450V—XXX'. The teacher was then tasked with testing the learner's recall of a list of pairs of words via a microphone. Upon hearing incorrect replies, the teacher was prompted by the experimenter (an actor) to administer shocks of increasing voltage, up to the potentially lethal 450V limit, despite the scripted cries of pain from the learner. Twenty-six of the volunteers—65 per cent—continued with the task until this conclusion, when the true nature of the experiment was revealed.

This experiment was inspired by the events of the Holocaust. Milgram wanted to explore the level to which ordinary people could be induced to harm their fellow humans at the request of an official or authority figure.

Milgram (1933–84) is also known for his 'small world' experiment, which tracked the progress of a package from a randomly selected individual in Omaha, Nebraska, or Wichita, Kansas, to a named person in Boston, Massachusetts. This is often cited as establishing an average of six links between any two people, but the results showed that only 64 out of 294 packages reached their destination.

In the early 1960s, University of Oklahoma professor William Lemmon attempted to establish a primate research centre on a farm. He theorized that previous ape studies had failed because they were too brief and began when the infant was too old. He planned a further study in humanizing chimpanzee behaviour.

LUCY THE HUMANIZED APE

MAURICE TEMERLIN

1966–77

University of Oklahoma psychology professor Maurice Temerlin and his wife, Jane, were recruited in 1966 to take part in an unusual investigation. It was less of an experiment and more of a behavioural study, during which they would rear a chimpanzee from birth as if she were their own offspring. The Temerlins committed to the project for the long haul. Jane collected the female chimpanzee, whom she named Lucy, from the circus where she had just been born and brought her into the family home. Over the next decade, the Temerlins raised Lucy as a human and taught her to dress, to use the toilet (with limited success) and to join them in their daily meals, eating, like they did, with a knife and fork. Lucy learned to make tea for guests and even to mix a gin and tonic, a beverage that she enjoyed before her evening meal. She was taught sign language and learned to communicate in two- or three-word sentences. As Lucy became sexually mature, she began to masturbate around the house, using a vacuum cleaner as a masturbation tool. Intrigued by her sexual responses, Temerlin provided her with a copy of *Playgirl* magazine. Lucy appeared to appreciate the naked men it depicted, but proved less responsive when introduced to male chimpanzees. It seemed that she had grown up to view herself as more human than ape.

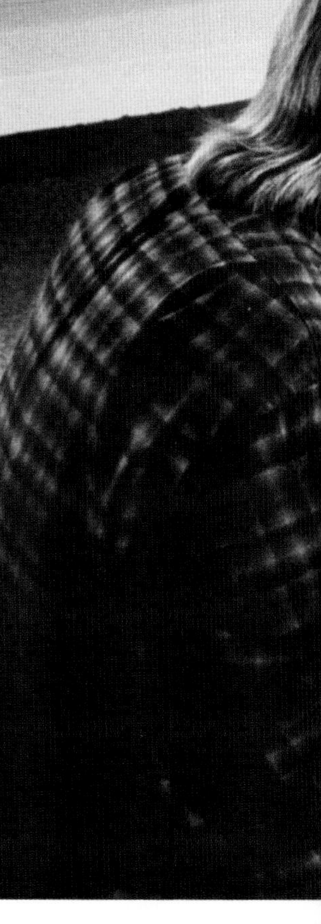

The professional life of Temerlin (1924–88) was profoundly impacted by his involvement in the Lucy study. Shortly after it ended, he resigned his position at the university and entered private practice. He wrote a number of psychology books with his wife.

Similar experiments involving chimpanzees were conducted around the same time, but these were concerned with language acquisition rather than socialization. All of them encountered problems when the chimpanzees reached maturity and were unable to live independently in human society. Such experiments fell out of favour and modern researchers began to focus on understanding chimpanzee behaviour instead of attempting to teach the animals to ape human traits, habits and communication.

The emotions she exhibits most clearly are affection, anger, fear, joy, tenderness, greed, jealousy . . . .

Lucy was taught basic sign language, although language learning was not the central concern of the study.

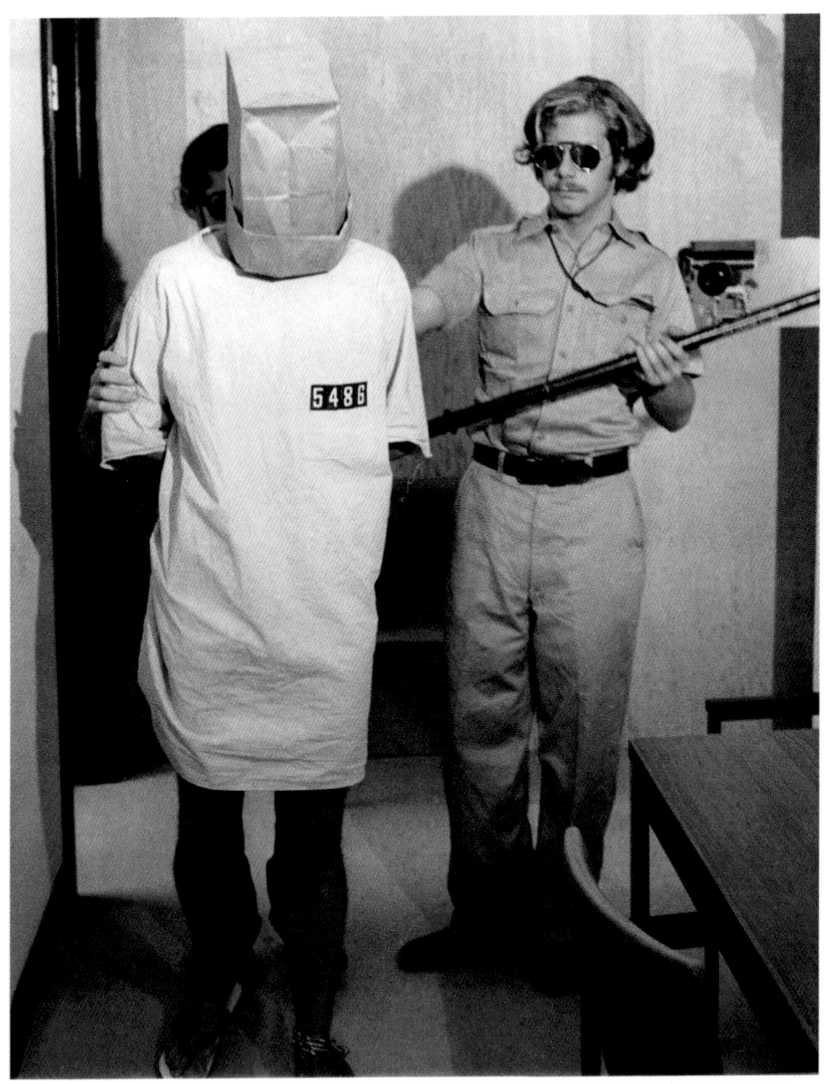

?

Zimbardo believed that his study not only demonstrated the importance of the psychology of the institution of prison, but also suggested the need for improved guard training. However, the experiment was criticized for its methodology and ethical considerations. One concern was that Zimbardo's deliberately vague briefing encouraged the guards to use psychological tactics to control the prisoners.

PRISON BEHAVIOUR
PHILIP ZIMBARDO
1971

In 1971 US psychologist Philip Zimbardo placed a newspaper advertisement asking for paid volunteers to take part in a study of prison life. All the applicants were assessed and twenty-four college students were invited o participate. These volunteers were then divided at ndom into 'prisoners' and 'guards'. Zimbardo briefed the uards, with vague instructions to maintain order and a ore specific directive on how to record their interaction ith the prisoners as well as the details of meals, work nd a recreation programme. An arrangement was made ith the local police, and the prisoners were collected nd taken through the procedure of arrest before being elivered to the simulated prison complex at Stanford niversity. After the first day, when the routines and tmosphere of the prison were established, antagonism etween the two groups developed rapidly as the guards ecame increasingly sadistic, dehumanizing the prisoners. outines rapidly broke down and the guards placed reater emphasis on forcing prisoners to recite rules that ney had drawn up. Conditions continued to deteriorate nd the guards went out of their way to break the morale f the inmates, including by inflicting punishments. Two risoners walked out and the experiment was brought to premature halt after only six days.

The early 1970s was a time of great social upheaval in the United States. In the wake of reports of prison guard brutality, Zimbardo became interested in the root cause of such violence and devised an experiment to examine if there were any inherent characteristics of prisoners and/or guards that might explain the abusive behaviour witnessed in prisons.

Zimbardo (1933–) has had a long successful career in psychology and has investigated many areas of the discipline. He became professor emeritus at Stanford University and also received the American Psychological Association Gold Medal for Lifetime Achievement in the Science of Psychology. In 2004 Zimbardo was called upon as an expert witness in the court martial of one of the soldiers in the Abu Ghraib prison abuse scandal.

In only a few days, our guards became sadistic and our prisoners became depressed and showed signs of extreme stress.

In order to promote feelings of anonymity and rank, uniforms were issued to all participants. Prisoners were given 'dresses' and numbers instead of names. The guards were provided with a khaki uniform, billy club and mirrored sunglasses.

Rosenhan (1929–2012) popularized the use of expert psychological opinion within the US court system, particularly in terms of jury selection and jury consultation. He served as president of the American Psychological Association and united his professional interests of law and psychology as director of the American Psychology-Law Society and president of the American Board of Forensic Psychology. During his very active academic career, he served on the faculties of Stanford and Princeton universities, as well as acting as a visiting lecturer at Oxford University, Tel Aviv University and the University of Western Australia.

It is clear that we cannot distinguish the sane from the insane in psychiatric hospitals. The consequences to patients hospitalized . . . seem undoubtedly counter-therapeutic.

DIAGNOSIS OF MENTAL ILLNESS

DAVID ROSENHAN

1973

In the early 1970s, US psychologist David Rosenhan began to enquire into the reliability of psychiatric diagnoses and how this might be tested experimentally. With seven of his associates, he decided to take on the role of a pseudo-patient. They all feigned auditory hallucinations in order to be admitted to various psychiatric hospitals across the United States. On admission, the pseudo-patients acted normally and explained to staff members that they felt fine and no longer heard voices. They complied with instructions and accepted, but did not swallow, their medication. All but one of the experimenters was diagnosed initially with schizophrenia, and when they convinced the staff that they were well enough to be discharged, all eight were certified as having schizophrenia 'in remission'. None of the hospitals detected that the pseudo-patients were not mentally ill. Rosenhan conducted a follow-up study in which he challenged another psychiatric institution to detect any pseudo-patients that he sent to them. The institution informed him that of the 194 patients who had been admitted, staff had identified a total of nineteen pseudo-patients. In fact, Rosenhan had not sent any.

In the mid 20th century, the methods employed to diagnose mental illness were highly subjective and unreliable. It was common for patients who were suffering from mental illness, or insanity as it was termed at the time, to be viewed as less than human. Many of the facilities were situated in historical, outdated buildings that were no longer fit for the purpose of modern care. Conditions were often harsh and cruel, and the treatments ineffective.

Rosenhan's experiment demonstrated the flaws in diagnosing mental illness: it was common for staff to over-diagnose such conditions, except when the pressure to do so was reversed, as in the case of his follow-up study. It also highlighted how a single diagnosis of mental illness can negatively affect subsequent assessments of a person's behaviour. Although there was some controversy over Rosenhan's experiments at the time—a number of psychiatrists contested the methodology—they forced many institutions to rethink their approaches to patient care and the diagnosis of mental illness.

St. Elizabeths Hospital in Washington, D.C., was one of the twelve institutions to which pseudo-patients were admitted.

Nim learns ASL with one of his teachers, Joyce Butler.

Terrace observed that although Nim was learning noun and verb signs, he often required prompting. When he reviewed the videotapes of Nim's interactions, he also noticed that the chimpanzee was not using the signs syntactically, and the significance and effect of word order or meaning were not being observed. Ultimately, Terrace concluded that, contrary to appearances, Nim was not acquiring language but was just 'a brilliant beggar', who had learned a series of imitations to perform when prompted in order to gain rewards.

LANGUAGE LEARNING IN CHIMPANZEES

HERBERT TERRACE

1973–77

In 1965 US psychologists Allen and Beatrix Gardner established an informal project at the University of Nevada, Reno, to teach a chimpanzee to communicate using American Sign Language (ASL). With researcher Roger Fouts, the couple interacted exclusively through sign language with their chimpanzee, named Washoe, until she learned about 150 signs. When she began to reach maturity at the age of five, the Gardners ended the project. Herbert Terrace led a similar experiment in the 1970s. He wanted to move beyond previous studies and test whether the instinct towards language, and most especially grammar, was innate in chimpanzees as it was thought to be in humans. Terrace's subject was a male chimpanzee, acquired shortly after birth. He was named Nim Chimpsky in reference to the linguist Noam Chomsky, who had identified language as being both innate in and exclusive to humans. Terrace exposed Nim to a variety of learning environments, from a family home to a more formal classroom setting, and provided him with several ASL instructors. He was monitored vigorously and most of his lessons were filmed for subsequent analysis. Over the course of the experiment, Terrace recorded more than 20,000 multi-sign sequences.

Attempts to teach apes to communicate using speech had limited success because the primate mouth and larynx struggle to form human sounds. A shift in focus to teach non-verbal methods was more successful.

US psychologist Terrace (1936–) received his doctorate in psychology at Harvard in 1961 and went on to specialize in cognition and intelligence among animals, with a particular emphasis on primates. He continued to work in the field of primate intelligence after the conclusion of Project Nim, moving his focus to rhesus monkeys. He is currently a professor at Colombia University Department of Psychology.

FREE WILL
BENJAMIN LIBET
1983

In 1983 US neurologist Benjamin Libet conducted an experiment to assess the time it took for a test subject to make a decision to press a button and for them to actually do it. In order to ascertain when the participant made the conscious decision to act, Libet attached electroencephalograph electrodes to the subject's head to monitor neural activity in the cortex. He then placed a modified oscilloscope in their field of vision. The oscilloscope displayed a dot that moved quickly in a circular motion, similar to the second hand of an analogue watch. The participant was asked to report the position of the dot at the time they made the decision to press the button. This enabled Libet to record when the conscious decision to move was made. He noticed that the activity in the brain—which indicated the brain was preparing muscles to move—spiked about 500 milliseconds before the hand pressed the button. However, the subject made note of the moving dot, and therefore the decision to move their hand, only 200 milliseconds before pressing the button. Libet conducted his experiment numerous times, but his results remained consistent, with a margin of error of only 50 milliseconds.

In the 1960s, German neuroscientists Hans Helmut Kornhuber and Lüder Deecke monitored electrical activity in the brain and discovered that an electrical potential can be detected in the organ before a person moves their finger. They termed this activity Bereitschaftspotential, or 'readiness potential', and in the decades that followed Libet began to investigate the extent to which readiness potential preceded not only bodily movement, but the decision to move itself.

During his wartime service, Libet (1916–2007) showed a strong interest in experimental proof when he was charged with assessing the effectiveness of three alternative boots to help combat the problem of trench foot—a feat that earned him an Award of Merit. He also received a Virtual Nobel Prize in 2003 from the Cognitive Psychology Unit of the University of Klagenfurt 'for his pioneering achievements in the experimental investigation of consciousness, initiation of action and free will'. The prize was formulated to recognize achievement in the field of psychology, a discipline not recognized by the Nobel Committee itself.

In his experiment, Libet monitored the activity of the brain's motor cortex through electroencephalograph electrodes, seen here in an angiogram (blood vessel X-ray).

The experiment demonstrated that an unconscious spike of brain activity preceded the conscious decision to act. This suggests that human consciousness lacks free will, or at least that brain activity occurs before conscious decision making. Libet's assessment of the results was that while the subconscious mind needed to prepare the body for movement, it was possible for these preparations to be countermanded by the conscious mind as an act of 'free won't' as opposed to 'free will'.

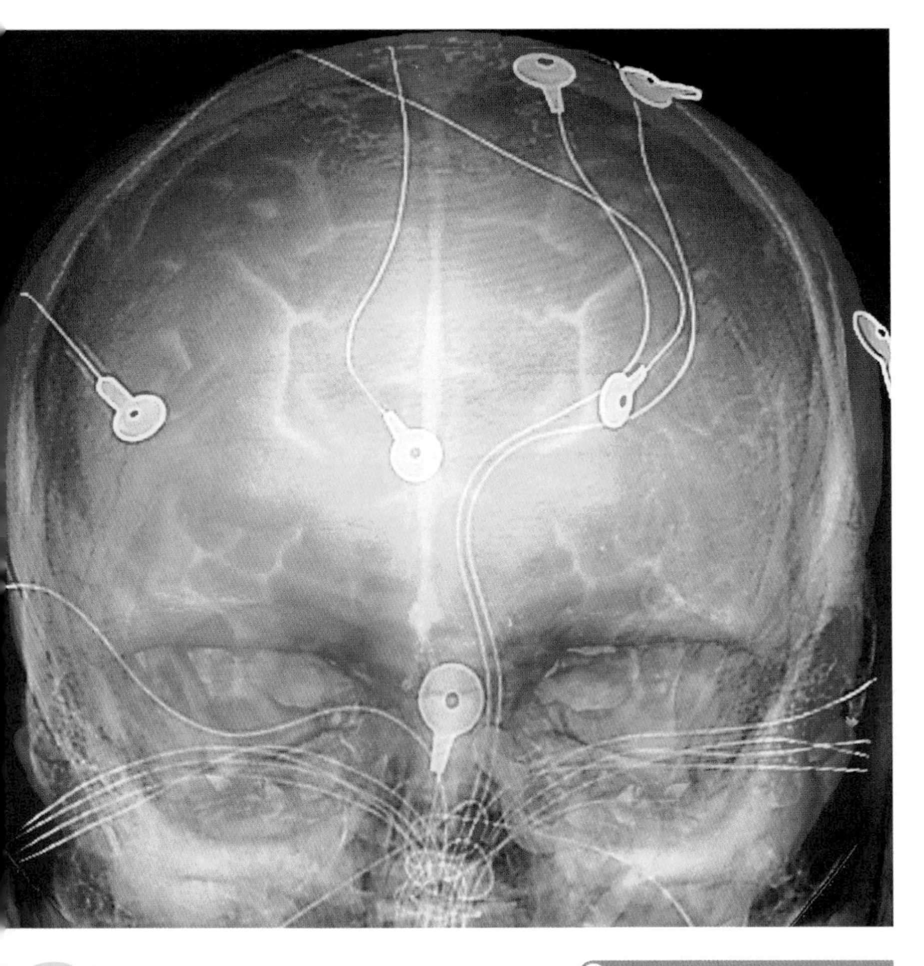

The finding that the volitional process is initiated unconsciously leads to the question: is there then any role for conscious will in the performance of a voluntary act?

Dan works in what is known as the 'Dan Lab' at the University of California, Berkeley. In recent years, she has focused her work on mammalian brain function, specifically the neural circuits controlling sleep and how the prefrontal cortex regulates the work of other areas of the brain.

To reconstruct movie scenes in an area large enough to contain recognizable objects, we pooled the responses of 177 cells.

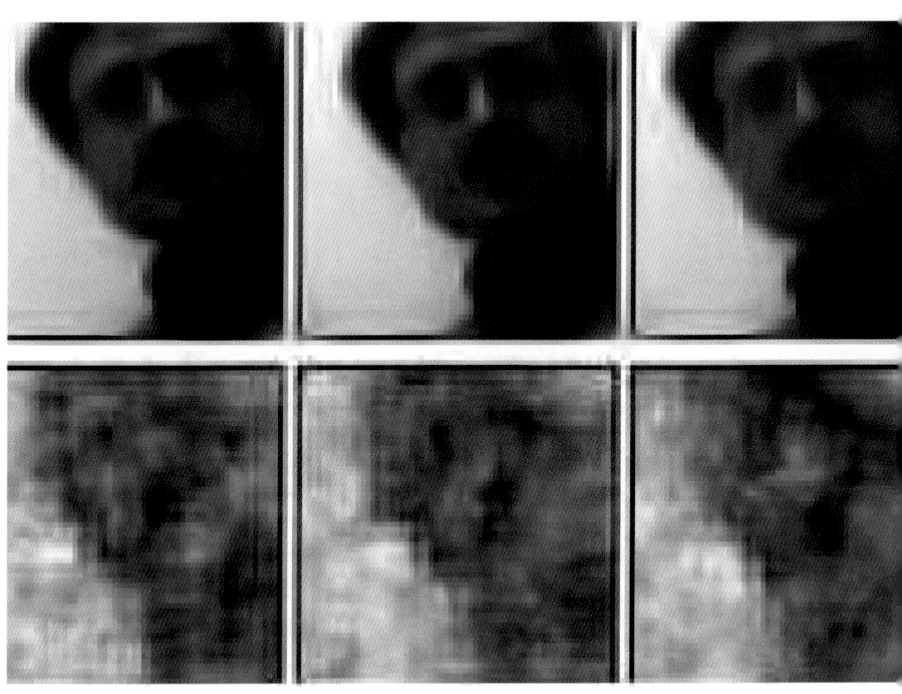

This experiment was the first direct demonstration that audiovisual images can be reconstructed from data collected from neuron activity in the brain. It showed how information from individual neuron activity can be used to reconstruct a more complete neural network. Over and above its value in further delineating brain function, Dan's experiment suggests intriguing possibilities for advances in technology in the field of human–machine interface. This could have implications not only for the creation of more effective and responsive prosthetics but also for other commercial applications, including eyeball cameras capable of transmitting moving pictures in real time.

The top row of images is from the actual source film shown to the cat; it depicts a man looking at the camera. The bottom row shows the corresponding images reconstructed from the cat's neural cells.

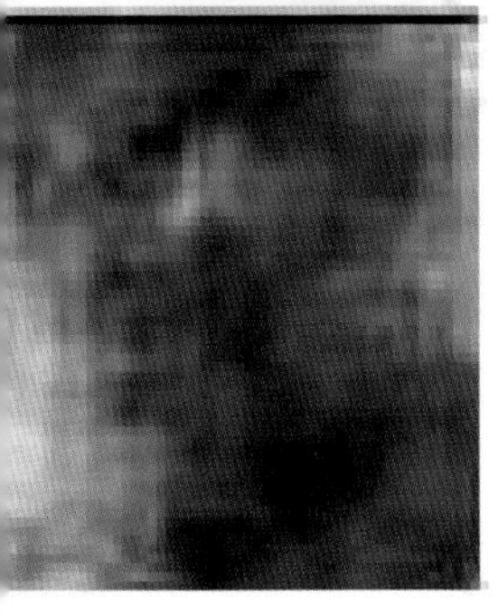

BRAIN–COMPUTER INTERFACE

YANG DAN

1999

Neurobiologist Yang Dan began her brain–computer interface experiment by anaesthetizing and paralysing a cat. The subject was given a tracheotomy in order to ensure that proper breathing was maintained and it was positioned within a frame that would hold it in place. Next, a local anaesthetic was administered to the cat's scalp and the top of its skull was removed, thereby providing access to its brain. The animal's eyes were dilated artificially and a television screen was placed in its field of vision. First, the screen displayed a video of white noise, followed by eight black-and-white film clips showing a variety of scenes. These ranged from footage of a woodland environment to a domestic indoor scene, complete with humans. As the cat watched the videos, the team placed a multi-electrode array on its lateral geniculate nucleus, the part of the brain that is responsible for processing visual data. While the cat viewed the white noise clip, Dan's team mapped 177 neural cells relating to its visual response out of possible thousands. Next, the animal watched the eight other films eight times each and its neural responses for each viewing were recorded. Finally, using a series of algorithms, the team employed a computer to reconstruct the videos based on data from the recordings of activity across the mapped cells. The resulting images were blurry, but they were recognizable renditions of those that appeared on the source videos.

The cat did not suffer pain or distress, and the experiment was approved by the Animal Care and Use Committee at UC Berkeley.

In the late 1990s, Dan and her colleagues began to explore various methods to identify and decode the neuronal firings in the lateral geniculate nucleus, the part of the brain that is responsible for interpreting visual data. Neurobiologists' understanding of how each part of the brain works is not yet complete and many areas of brain function remain poorly understood.

In the 1970s and 1980s, a number of psychologists conducted a series of studies on the role that attention plays in human perception. The formats varied, but most of the experiments employed videos to test volunteers' ability to perceive events to which they were not paying full attention.

CONCENTRATION AND PERCEPTION

DAN SIMONS AND CHRISTOPHER CHABRIS

1999

The perception experiment devised by US psychologists Dan Simons and Christopher Chabris centres on a video in which two teams of people—one wearing white shirts, the other wearing black shirts—play with a basketball. Each team has its own ball, which they dribble and pass between their teammates. The viewer is tasked with keeping a tally of the number of times that the basketball is passed between the people wearing white shirts, thereby ignoring any passes made between those wearing black shirts. The two teams pass their respective balls for about a minute. At the end of that time, it is revealed that the team in white performed fifteen passes. As the viewer is congratulating themselves on the acuity of their vision and their ability to count correctly, they are asked if they also saw the gorilla that appears for nine seconds in the middle of the video clip. The video is rewound and replayed, and sure enough there is a nine-second section during which someone wearing an unconvincing gorilla suit slowly saunters through both teams as they pass their basketballs. The gorilla pauses to face the camera and beats its chest before walking off. The viewer did not notice it the first time simply because they had been told to concentrate on something else and specifically to filter out people wearing dark colours, eg the gorilla.

Simons (1969–) and Chabris (1966–) are active academics, teaching psychology at the University of Illinois and Union College, New York, respectively. They shared the Ig Nobel Prize for Psychology in 2004, and in 2010 published their research in a book titled *The Invisible Gorilla*.

Simons and Chabris demonstrated how at least 50 per cent of people failed to see an unexpected event when they were asked to monitor something else. However, it appears that sustained inattentional blindness is even greater in professionals who are engaged in a familiar task. In a study in 2013, 83 per cent of radiologists, who routinely review slides in search of cancerous nodules, failed to spot that an image of a gorilla—forty-eight times the size of a standard nodule—had been superimposed onto one of the slides they were asked to check.

The 'gorilla' appears obvious in this sequence of stills, but, when viewed as part of the experiment, it frequently goes unnoticed.

In *An Evening of Wonders*, British illusionist Derren Brown performed a variation of the gorilla experiment to illustrate change blindness.

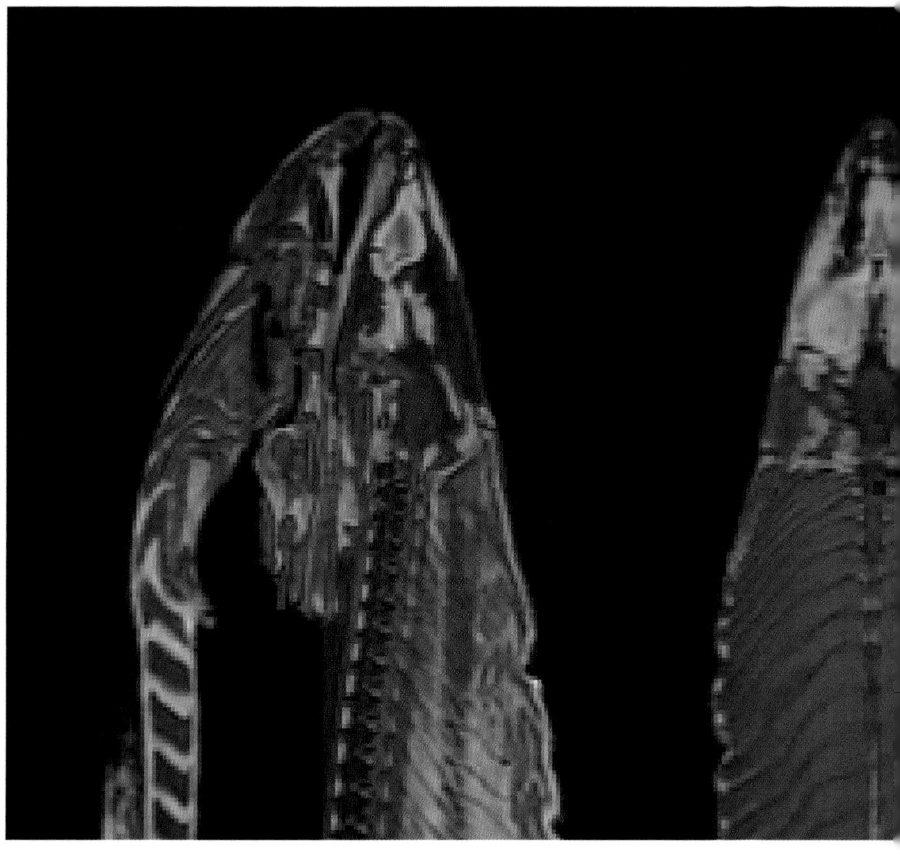

The results collected by Bennett and his colleagues demonstrate the danger of over-interpreting data and of failing to identify false positives. This can be a particular problem with fMRI scans in which the failure to correct the threshold values before a scan can lead to unwanted, random and false signals drowning out or disguising the importance of the actual accurate data. Bennett's group also conducted a meta-analysis of major journal articles that included fMRI data. In 25 to 40 per cent of the papers, they found that the threshold values had not been corrected properly, which brought the results into doubt. Fortunately, a follow-up study in 2011 demonstrated fewer neuroscientists were making the same mistake.

An image of the salmon's fMRI scan was circulated at a Human Brain Mapping conference in San Francisco, where it provoked much discussion.

Functional magnetic resonance imaging (fMRI) measures brain activity by tracking changes in blood flow to parts of the brain. It is widely used in modern medicine and fMRI results are relied upon to deliver diagnostic information for a range of neurological conditions.

fMRI AND NEUROSCIENCE

CRAIG BENNETT

2005

A somewhat unorthodox experiment began in 2005 when neuroscientist Craig Bennett and his colleagues placed an unusual test subject in an fMRI scanner. The results of the scan did not mean much to the subject—it was an Atlantic salmon and was already dead—but they did have implications for the future of fMRI imaging. During the experiment, the salmon was shown a series of photographs that depicted humans in various social situations. Like a human patient in similar circumstances, the fish was then invited to describe the emotional state of the people in the images. Using the machine's standard presets, the fMRI produced an image of the fish with 3D pixels, or 'voxels', showing activity in the salmon's brain. This scan result indicated that the subject had experienced an emotional response to the pictures. Of course, the dead fish was not experiencing anything at all, but by chance the scanner had generated a false signal that could easily be misinterpreted.

We found some [signals] that were significant that just happened to be in the fish's brain. And if I were a ridiculous researcher, I'd say, 'A dead salmon perceiving humans can tell their emotional state.'

Bennett (1980–) and his team jointly received the Ig Nobel Prize for Neuroscience in 2012 for 'demonstrating that brain researchers, by using complicated instruments and simple statistics, can see meaningful brain activity anywhere—even in a dead salmon.'

Physician and cosmonaut Morukov (1950–2015) knew first hand about the practical concerns of space flight, because he had spent nearly twelve days in space in 2000 on the NASA Space Shuttle mission STS-106 to the International Space Station. He became the senior researcher in metabolism and immunology at the IBMP.

MARS MISSION SIMULATION

IBMP AND ESA

2010–11

On 3 June 2010, an all-male international crew embarked upon a 520-day simulation of a manned mission to Mars, under the direction of project leader Boris Morukov. It was a collaboration between the Russian Institute for Biomedical Problems (IBMP) and the European Space Agency (ESA). The crew's home was a small, purpose-built, sealed facility located on the IBMP site in Moscow. It consisted of three spacecraft modules—stocked with food, medical supplies and other equipment necessary for a long trip to Mars—one lander module and a simulation of the Martian landscape. The experiment was divided into phases to mimic the round trip and it included a simulated landing on Mars for some of the crew. Communications with the outside world were limited, and an artificial communication lag was introduced to recreate the conditions that would be faced on a real mission. This added to the sense of isolation that was the focus of the experiment. It also affected the sleep patterns of the crew. Some slept longer, taking naps during the day, while another did not sleep well at all, which impacted on his concentration and alertness.

Once the door closed on the Mars 500 mission, the six-man crew was sealed in the 550-cubic-metre (19,423 cu ft) module for 520 days.

A round trip to the Moon would take about a week, but Mars is a lot further away and a return journey could take between one-and-a-half and two-and-a-half years. Such a mission would involve lengthy periods of isolation for the crew and in order to prepare for this scientists have planned a number of long-term space simulations; Mars 500 is the longest to date.

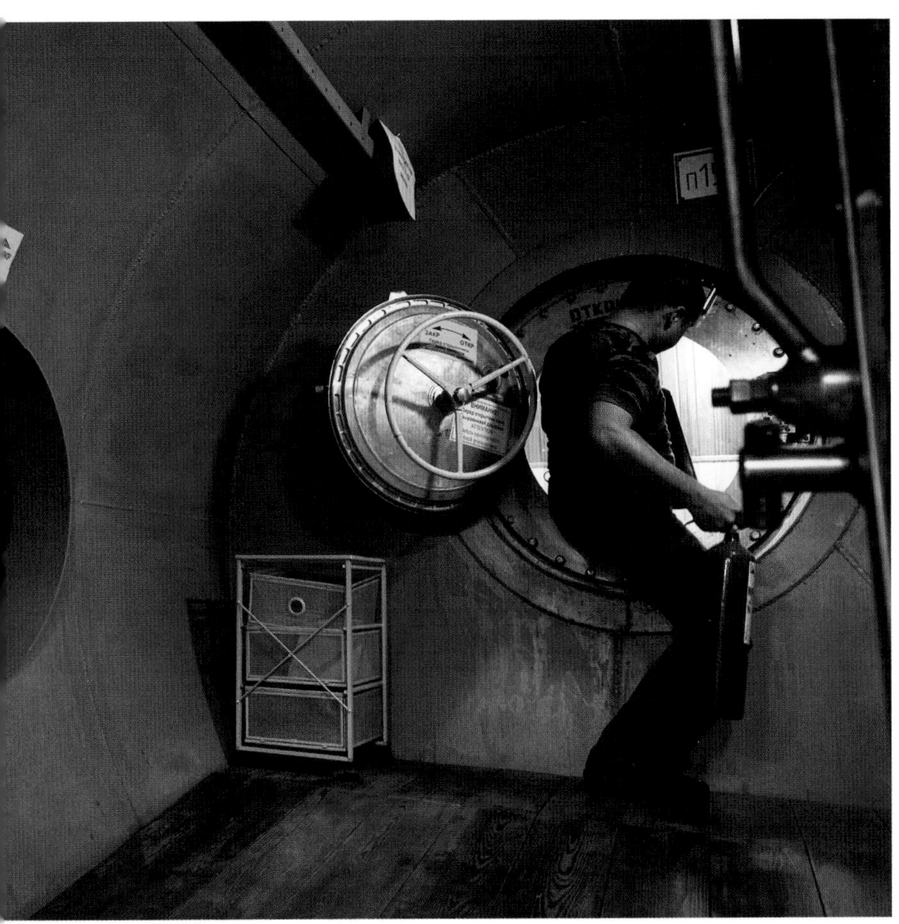

"

You have to adapt to not seeing the sun, not seeing grass . . . I think in space it is easier—there's a window, you can see the stars.

SUKHROB KAMOLOV, MISSION PHYSICIAN

?

The scientists behind Mars 500 assessed the performance of the crew as a single unit and noted how well they worked together. There was an absence of conflict based on cultural or linguistic differences, despite the international nature of the crew. Further data from the experiment has led to follow-up research on topics including the effects of exercise on cognitive performance. NASA has begun conducting its own Mars-focused crew isolation experiment in Hawaii, although no manned Mars mission is planned before 2030.

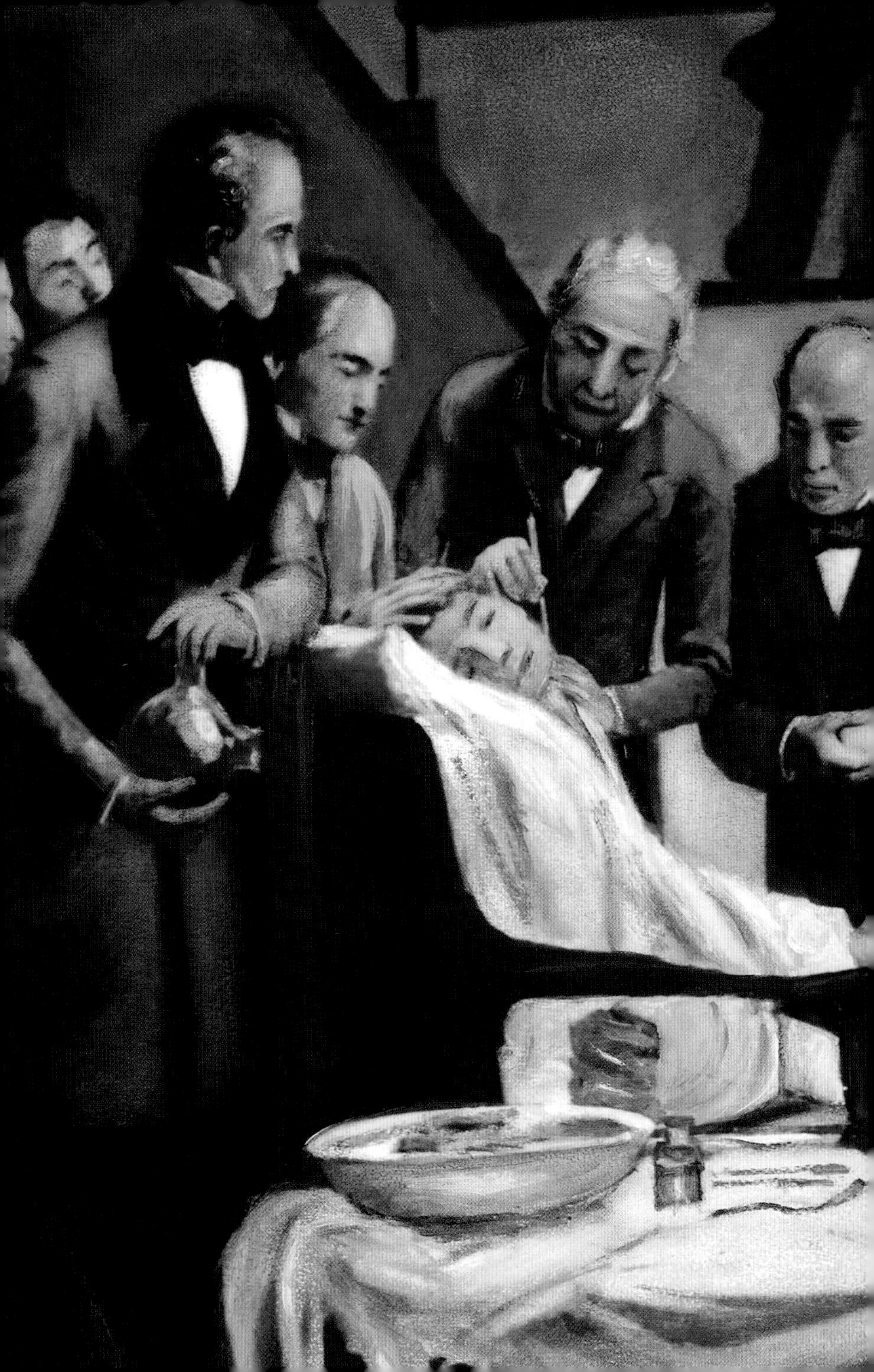

CHAPTER THREE
SOCIETY

A great variety of experiments has been conducted under the premise of benefiting society as a whole, conceived largely in either concrete or abstract terms of knowledge gained. For example, Carl Linnaeus worked to systematically group together all living things and Lawrence and William Bragg achieved a breakthrough in X-ray crystallography that was later used to facilitate the manufacture of organic compounds and to discover the structure of DNA. Some experiments, such as Vijay Pande's ongoing Folding@ home project, continue to provide science for the betterment of humanity, whereas others, such as the syphilis study in Tuskegee, resulted in an undeniably negative effect on its participants. In the case of James Vicary's subliminal advertising experiment, it is likely that there was no tangible effect at all.

‹ Ether as a General Anaesthetic (see p. 102)

In *Systema Naturae* (1735), Linnaeus aimed to impose a structural order on all things animal, vegetable and mineral. Although minerals are no longer classified in the same way as life, the book had a great impact on the field of biology and Linnaeus's work still underlies how we classify life to this day. He also created the binomial (two name) classification system of scientific names that facilitates modern taxonomy.

LINNAEAN TAXONOMY

CARL LINNAEUS

1735

Although scientists had begun to classify plants and animals more systematically by the 18th century, no single system had been adopted. In 1735 Swedish botanist Carl Linnaeus moved to the Netherlands, where e met a number of botanists and discussed the virtues f a classification system that he had been developing. Dutch botanist named Johannes Burman, who was ʼorking on his own catalogue of the flora of Ceylon ›resent-day Sri Lanka), challenged Linnaeus to classify a ıystery plant from a sample that Burman would provide. he sample consisted of dried leaves and withered owers. Linnaeus placed one of the flowers in his mouth ı an attempt to moisten it and revive its original shape. ooking again at the plant, he proclaimed the sample to e from a laurel tree closely related to the bay laurel. owever, Burman informed him that it was, in fact, from cinnamon tree. Linnaeus agreed, but further explained ow his classification worked to group together similar lants with similar attributes. He won Burman around to ccept his scheme, which he later developed to include a imple yet effective naming system that consisted of two arts: the first to identify the genus, or group, and the econd the species of each life form. Consequently, the innamon tree is known as *Cinnamomum verum* and is laced within the wider order of laurel trees, Laurales.

In *c.* 320 BCE, ancient Greek botanist Theophrastus began to classify plants mainly by their size, fitting them into four groups: trees, bushes, shrubs and herbs. Many centuries later, plants were commonly categorized for their uses to humanity, as food, medicine, an aromatic or a building material.

Linnaeus (1707–78) studied botany and medicine, and travelled widely in Sweden, collecting plant samples. In addition to *Systema Naturae*, he penned a number of books on botany, many of which focused on the flora of Sweden. Linnaeus is also credited with inverting the numerical values of the Celsius scale, invented by Anders Celsius, ensuring that zero degrees marked the freezing point of water and one hundred degrees the boiling point.

No one has been a greater botanicus or zoologist. . . . No one has more completely changed a whole science and initiated a new epoch.

This page from the *Universal Technological Dictionary* (1823) by George Crabb depicts mammals and reptiles classified according to the taxonomy of Linnaeus.

German scientist Ritter (1776–1810) is credited with the discovery of ultraviolet light. Inspired by the existence of infrared light beyond the red end of the visible spectrum, he reasoned that there must be another invisible light at the other end to balance it. For his experiment, he used a prism to create a spectrum and then exposed each colour within it to silver chloride, which he knew darkened as a reaction to natural light. He noticed that the silver chloride became darker more quickly when placed at the extreme end of the spectrum beyond violet.

Volta's diagrammatic drawing of his first voltaic pile was published in the Royal Society's journal *Philosophical Transactions* in 1800.

Ritter did not hide the fact that he was a great believer in the occult, and for this reason many of his contemporaries disregarded him as a serious scientist.

EFFECTS OF ELECTRICITY

JOHANN WILHELM RITTER

c. 1800

Johann Wilhelm Ritter's relatively short career spanned the early years of the 19th century and included some unusual experiments. A number of these involved his own version of a voltaic pile: a simple early battery newly invented by Italian physicist and chemist Alessandro Volta. Ritter experimented with the composition of his pile, favouring the use of copper disc as a cheaper alternative to the silver ones that were sometimes employed, interspersed with brine-soaked cloth. As he applied the pile's electricity to his own body in order to test its effects on his senses, Ritter increased the height, and therefore voltage, of his pile to as many as one hundred discs to maintain the current against the resistance of his flesh. In his results, he recorded completely different reactions in his sense organs—his tongue (acidic taste), nostrils (sneezing sensation) and eyes (strange vision)—when he connected them to the pile. However, some of Ritter's observations, such as feelings of warmth and looseness or cold and stiffness in his muscles when touching wires attached to the pile, could not subsequently be confirmed or reproduced. He even applied electricity to his own genitals, stimulating himself to climax. Ritter repeated these experiments with such frequency and increasing current that they impacted on his health and he died at the age of thirty-three.

In the 1800s, electricity was a poorly understood but exciting phenomenon for scientists. Influenced by Italian physician Luigi Galvani, who had experimented with electricity on the nerves of frogs' legs, Ritter was drawn to investigate the effects of electric shocks on various parts of the human anatomy.

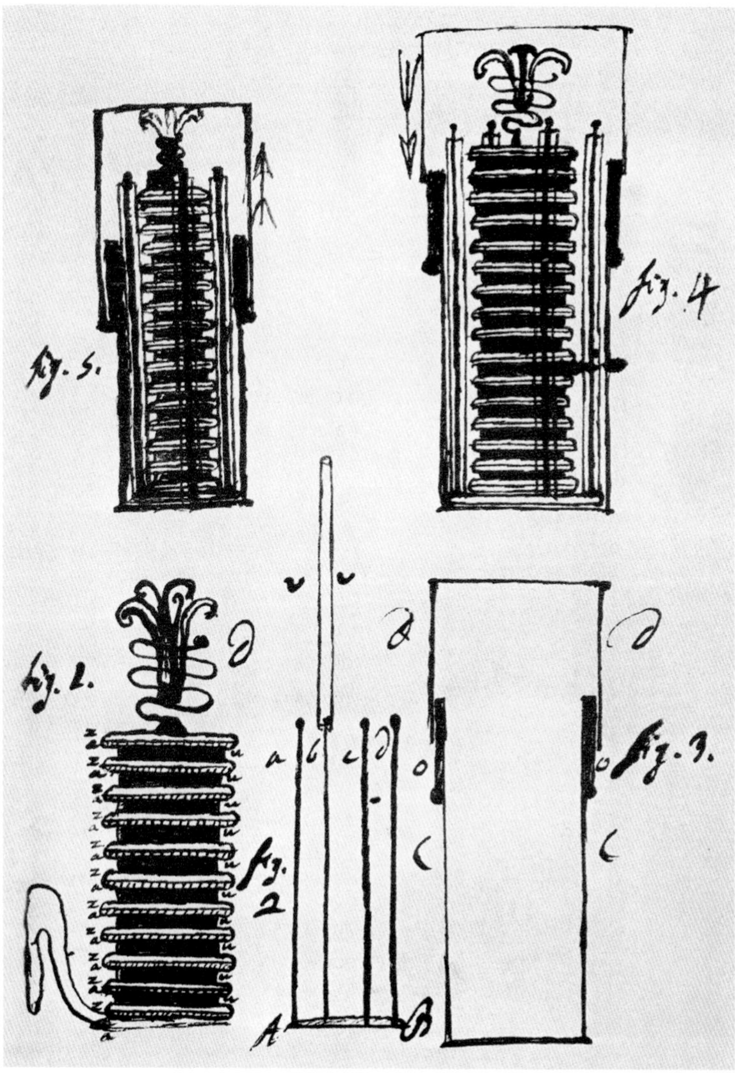

Ritter's passion for self-experimentation served as a warning to scientists against the dangers of excess. He often failed to fully grasp the science behind the phenomenon he was observing and his eccentric behaviour did not help his reputation. However, after the experiments that he conducted on his own sensory organs, he went on to invent the dry-cell battery in 1802 and an electrical storage battery in 1803.

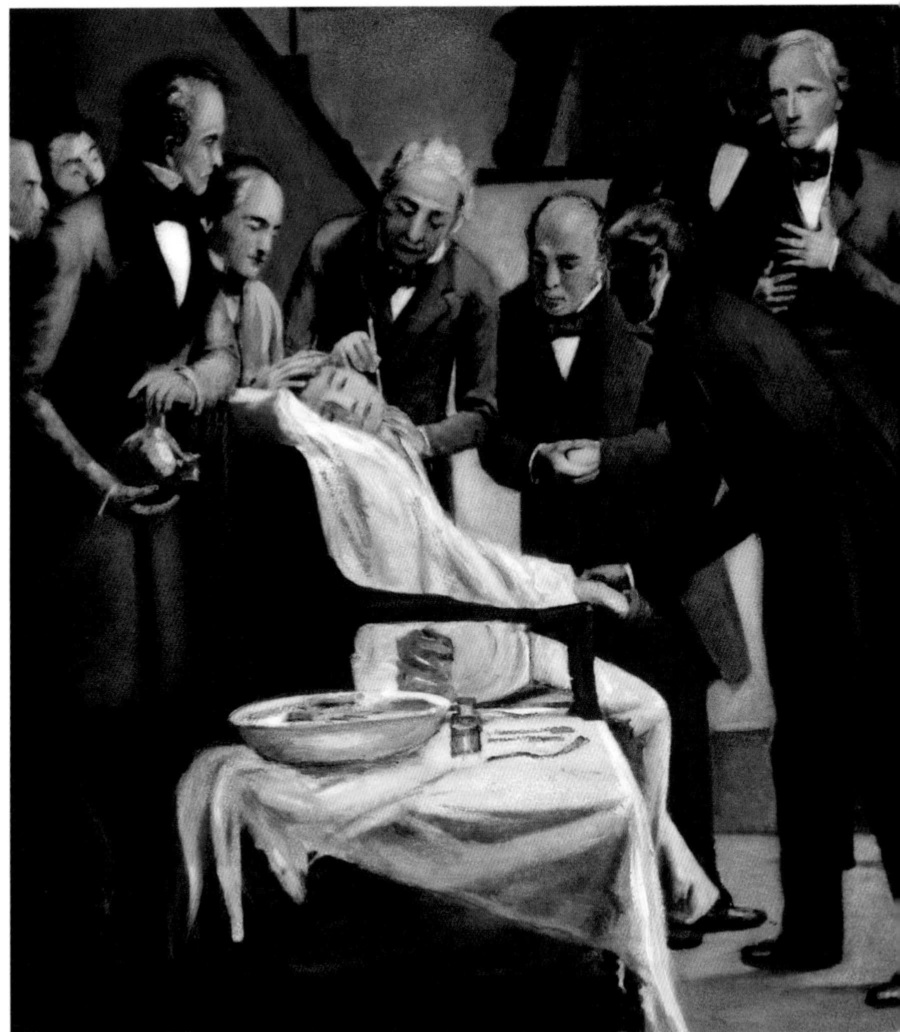

?

The demonstration of diethyl ether as an effective means of pain relief heralded the age of anaesthesia in medicine, increasing patient comfort and reducing some of the complications of surgery. Although the compound's popularity for medicinal use was relatively short lived, due in part to its flammability and the relative slowness of its pain-relieving effects to manifest in patients, its efficacy helped to speed up the adoption of alternative compounds, such as chloroform.

In the 1840s, the compound diethyl ether was consumed by physicians, medical students and the curious public at events dubbed 'ether frolics'. It was popular as a recreational intoxicant and an alternative to alcohol. But did it have a greater medical purpose, and could it be used in the operating theatre?

US dentist and medical student William Morton was determined to find an effective and reliable method of pain relief for use during surgery. He tried everything from brandy to laudanum and opium in a series of experiments carried out on himself and his dental patients. After a long process of trial and error, he hit upon the idea of inhaling diethyl ether as a means of administering an anaesthetic. Morton first tested his theory during a tooth extraction on 30 September 1846, and the patient reported no pain. When news of this success reached the staff at Massachusetts General Hospital, noted surgeon Henry Bigelow arranged the first public demonstration of the method on 16 October 1846, during which Morton administered the compound before surgeon John Collins Warren operated on a tumour on the patient's neck. Although this first demonstration did not quite go to plan—the patient complained of considerable pain 'as though the skin had been scratched with a hoe'—Morton tried again the next day with greater success. This time, the patient—a woman with a 'fatty tumour of considerable size' on her arm—reported feeling no pain whatsoever; she did not even recall the operation itself.

Morton (1819–68) spent much of his life chasing recognition and recompense for his pioneering work with ether. Initially, he tried to conceal the composition of the wonder drug, which he called 'letheon', and then he attempted to patent his methods. Neither venture was successful.

Snow (1813–58) took an interdisciplinary approach to his medical practice and received training in the fields of surgery and pharmacy before becoming a qualified doctor. In addition to his work on epidemiology and the causes of cholera, he experimented with the use of ether and chloroform as an anaesthetic. He investigated the administration of the substances in controlled doses, eventually making them safer and more effective. He also helped to popularize the use of chloroform in obstetrics when he administered it to Queen Victoria during the birth of her last two children in 1853 and 1857.

EPIDEMIOLOGY JOHN SNOW 1854

In the summer of 1854, Soho in London was gripped by a violent outbreak of cholera. Local doctor John Snow suspected that water contaminated with sewage was the cause. With the aid of a local parish priest, he went from door to door to collect information on the date and location of the cholera deaths and charted the results on a map of the area. He noted that the deaths were clustered largely around the Broad Street water pump, but his data also contained some anomalies. There were surprisingly few deaths in the local workhouse and none among the workers of a nearby brewery, whereas a woman from Hampstead had died from the disease. Further investigation revealed that the workhouse had its own well, and that the brewery workers had access to free beer, so neither used the Broad Street pump. The woman from Hampstead, on the other hand, had her drinking water brought to her specifically from the pump because she thought that it tasted better. Using this data, Snow convinced the local authorities to remove the pump handle and disable the well, thereby decreasing the number of cholera deaths.

This map shows where the various victims of the Soho cholera outbreak of 1854 lived, with a large cluster of cases located close to the Broad Street water pump.

By the mid 19th century, Asiatic cholera had become a global disease, but its cause and exact means of transmission remained poorly understood. Miasmas (foul-smelling air) and the dried discharge from victims were widely believed to be at fault, but Snow had another theory. It was based on his observation of the London cholera outbreak of 1848, in which the areas hardest hit were the ones that obtained their drinking water from the stretches of the River Thames that were most contaminated by human sewage.

Snow spent much of his life teetotal, but today a public house bearing his name is located near the Broad Street pump.

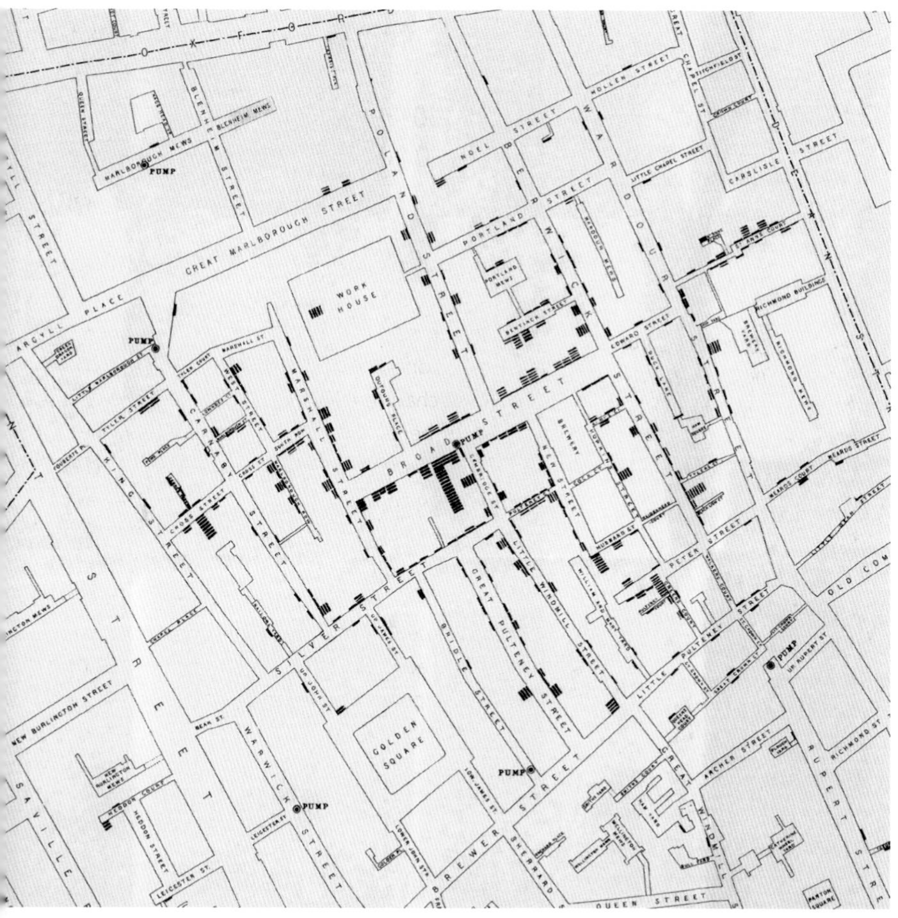

Today, Snow is revered as one of the fathers of the science of epidemiology for his systematic study of the London cholera outbreak. However, despite his evidence, doctors and politicians were reluctant to admit that the city's drinking water could be dangerously polluted with sewage. It was not until a further outbreak in London in 1866 that fellow doctor William Farr drew on Snow's work and concluded that cholera was indeed transmissible through contaminated drinking water.

"

The most terrible outbreak of cholera which ever occurred in this kingdom is probably that which took place in Broad Street . . . a few weeks ago.

Hofmann was protective of Perkin, and when he was told of the young man's intention to abandon his studies and enter into industry, he attempted to dissuade Perkin from this plan.

Organic chemistry was poorly understood in the mid 19th century, and mauveine was the first synthetic organic chemical dye to be discovered. Also known as aniline purple, the dye sparked a fashion craze for mauveine-dyed goods, and at the Royal Exhibition in 1862, Queen Victoria wore a silk gown dyed with Perkin's mauve. Perkin's unintentional discovery of the compound prompted a rapid expansion of both the synthetic dye industry and the wider large-scale chemicals industry. He went on to discover several other dyes, including Britannia Violet and Perkin's Green. His ongoing research in organic chemistry also produced the 'Perkin Reaction,' which established the basis for the synthetic perfume industry.

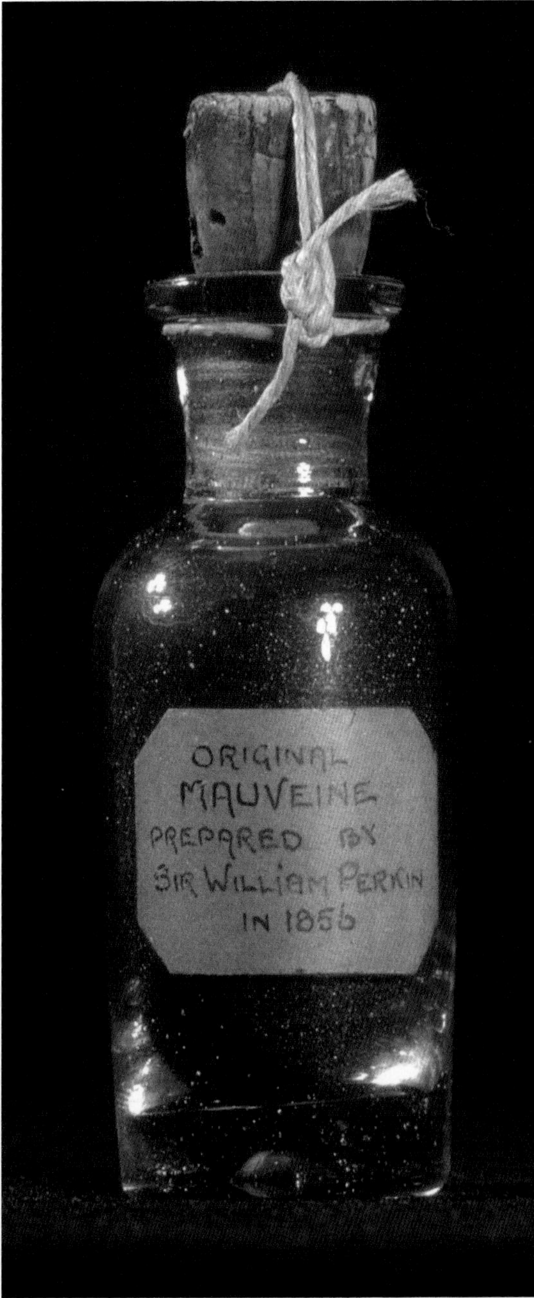

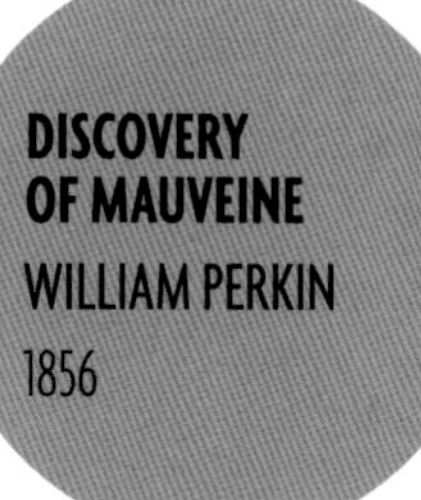

DISCOVERY OF MAUVEINE

WILLIAM PERKIN

1856

There are several accounts of British chemist William Perkin's experiment, and the details are variable. However, most sources agree that over the Easter period of 1856, when Perkin was only eighteen ears of age, he accidentally discovered a synthetic purple lye while working on German chemist August Hofmann's roject to synthesize quinine from coal tar products. erkin added potassium dichromate to a flask containing small impure quantity of the colourless, smelly, coal ar compound aniline. This produced a black sludge that eemed unremarkable until he poured methylated spirit nto the flask to remove the sludge. Perkin then noticed hat the sludge dissolved in the alcohol and the resultant olution turned a purplish colour. Further experiments ndicated that the substance would dye silk and that the olour would remain stable. Samples of the dyed silk and he dye itself were sent to the dye works of J. Pullar & ons in Perth, Scotland. They expressed a keen interest, xplaining that the samples appeared to be more durable han any of the then-existing means of producing a imilar colour. Perkin went on to patent his dye, which ame to be known as mauveine, in August 1856. With he financial support of his family, he set up a factory to nanufacture the dye on an industrial scale.

Quinine was known to be an effective treatment for malaria, but it was difficult to obtain. Hofmann postulated that it might be possible to synthesize quinine using elements from other compounds. Although he was mistaken, his idea of using coal tar, a by-product of the coal gas industry, was a plausible starting point at the time.

Following others would result in doing about as well as others; but the desire should be to excel and do much better than others.

Perkin (1838–1907) was born in London, the son of a prosperous carpenter. He attended the City of London School, where he developed a love of science, ultimately moving to the Royal College of Chemistry in 1853. There, Perkin became an assistant to Hofmann and participated in the latter's coal tar experiments. After his discovery of mauveine, Perkin moved into industry, where he enjoyed great success and profit. Once retired, he continued his research in the field of organic chemistry.

Although Perkin first patented his dye in 1856, it is thought that his early samples were not manufactured until around eight years later.

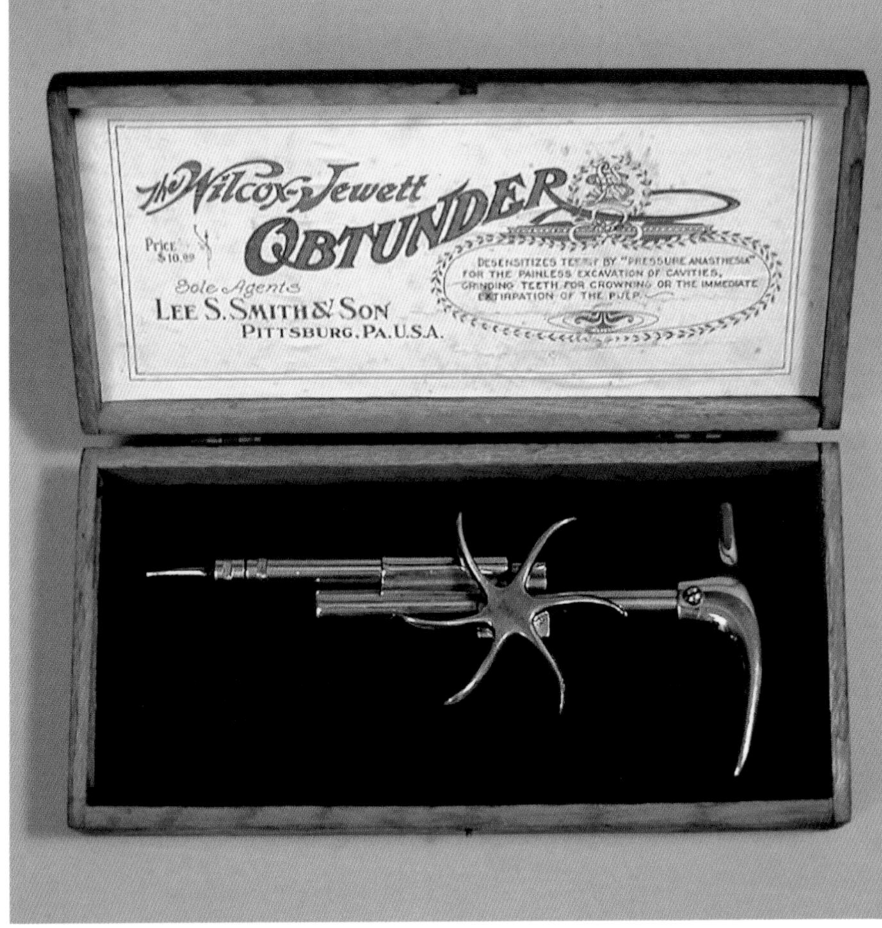

In early 20th-century dental practice, cocaine was administered using devices such as the Wilcox-Jewett Obtunder.

?

As a result of his experiments, Koller and his colleagues began using cocaine solutions of between 2 and 5 per cent both to alleviate the pain of patients' eye complaints and, more significantly, as a means of anaesthetizing patients during eye operations. The efficacy and usefulness of local anaesthetic spread through the medical community and became particularly popular in dentistry. However, in the early 20th century, cocaine fell out of favour due to its addictive and euphoric properties. Today's modern local anaesthetics are more effective and less addictive.

COCAINE AS A LOCAL ANAESTHETIC
KARL KOLLER
1884

Austrian ophthalmologist Karl Koller's series of experiments investigating the anaesthetic virtues of cocaine were not for the squeamish. He used a variety of animals as test subjects, including guinea pigs, rabbits and dogs, nd anaesthetized their corneas with a few drops of a per cent cocaine solution, administered directly into ach animal's eye. Koller found that he could touch an naesthetized animal's cornea with a pinhead without ausing the eyelids to close nor the head to jerk back reflex. Electric currents and poking the cornea with pencil produced no reactions either. Koller also xperimented on himself and his colleagues, using the ame dosage of cocaine drops and the head of a pin. He eported a 'slight burning sensation' that lasted a minute hen the drug was administered, followed by a lack of ain or sensation when the pinhead was pressed against he cornea, even when he pushed it hard enough to ause a temporary dimple. Koller reported that complete naesthetization lasted only about ten minutes, with ormal reaction returning slowly over a couple of hours. e also noted that the effect of the drug focused on the ocation where it was applied and that repeated application f the drug could increase the period of its efficacy.

Although ether and chloroform were popular in the field of pain relief in the mid 19th century, there remained a number of operations in which these anaesthetics were not effective. In optic surgery, involuntary reflex motions of the eye were caused by the slightest stimuli. Coca leaves had long been known for their pain-relieving properties and scientists were keen to explore new uses for a potential wonder drug.

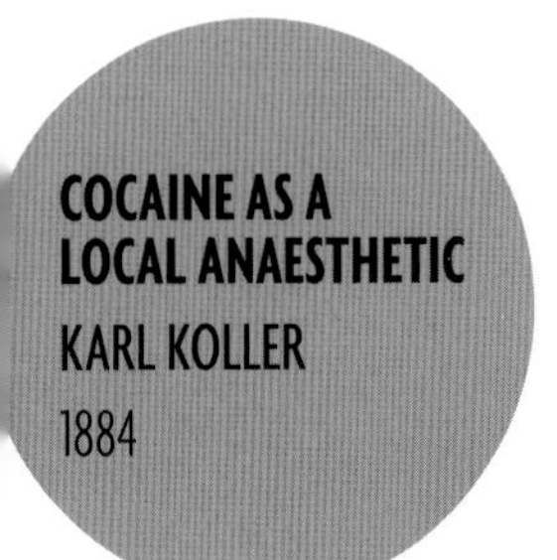

Koller (1857–1944) was a contemporary and friend of Sigmund Freud, and the pair shared an enthusiasm for cocaine. However, Koller was the first to publish a paper on the drug's potential use in medicine with the presentation he delivered to the convention of German ocularists in mid September 1884. Over the course of his lifetime, Koller gained significant recognition for his work as an ophthalmologist, and he received the Lucien Howe Medal in 1922 from the American Ophthalmological Society for his achievements in this field. The American Society of Regional Anesthesia and Pain Medicine awards an annual research grant named after Koller, and has done so since 1986.

The anaesthetic effect is pre-eminently a local one . . . it is stronger on those places to which the solution has been directly applied.

Tesla's work inspired engineers and investors, and his invention of the AC induction motor helped to spread the use of electricity in the United States. Indeed, his forecasts for future technology were often visionary. However, although his magnifying transmitter experiment was visually stunning, it highlighted the fact that impressive displays are no substitute for methodical and peer-reviewed experimental science.

At Tesla's request, the photographer enhanced this image to imply that the person reading and the sparks occurred simultaneously.

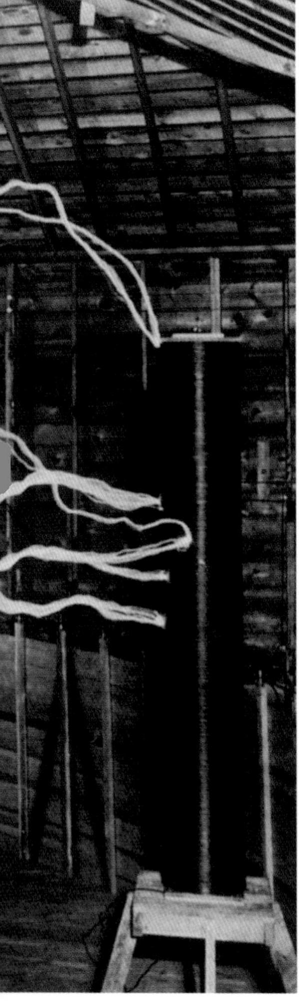

The electricity market of the 1880s was dominated by the war of the currents: Thomas Edison's company promoted direct current (DC) and rival companies preferred the more efficient alternating current (AC). Ultimately, Edison's firm adopted AC and entered into a merger to become General Electric.

MAGNIFYING TRANSMITTER

NIKOLA TESLA

1899

In 1899 Nikola Tesla relocated to a purpose-built laboratory in Colorado Springs, Colorado, where he had access to unlimited free electricity from the local power company. There, he built a magnifying transmitter, a more powerful and more advanced version of the Tesla coil that he invented in 1891 to allow the wireless transfer of electricity. This new device could emit streams of electricity up to 9 metres (30 ft) in length and was designed to transmit massive amounts of electricity wirelessly through the ground to power nearby devices. He also hoped to use a similar electrical device to transmit communication signals around the world. Tesla's claim that his transmitter successfully illuminated lamps on a loop of wire around the laboratory using electricity sent through the ground was difficult to prove because he was reluctant to give public demonstrations.

At Colorado Springs, Tesla detected a regular bleeping signal of unknown origin. He became convinced that it represented some kind of interplanetary message.

Serbian-American inventor Tesla (1856–1943) was very successful and obtained hundreds of patents during his lifetime. The SI unit of magnetic flux density, a crater on the Moon and Belgrade's international airport all bear his name. He even explored the potential of solar energy and tidal power.

In the late 19th century, chemists in Germany recognized that they needed a new source of nitrogen-rich fertilizers to ensure they could produce enough food to support their growing population. Haber's solution was to develop a process that turned the nitrogen in the air into a form that could be used in fertilizers.

PRODUCING AMMONIA

FRITZ HABER

1909

The problem that Fritz Haber faced in his attempts to convert nitrogen gas, which makes up 78 per cent of the air in the atmosphere, is that it is very stable and very difficult to break down into an element that will bond with others. However, Haber was methodical and set about examining all the ways in which atmospheric nitrogen could be transformed, finding that most methods produced yields far too low to justify the high energy costs. Drawing on the work of British chemists William Ramsay and Sidney Young, who detected trace amounts of the nitrogen product ammonia during their experiments on the decomposition of nitrogen at high temperatures, Haber found that he could produce ammonia in larger amounts by heating nitrogen and hydrogen at high temperatures, up to 1,000°C (1,832°F), and using iron and other metals as catalysts. Through an ongoing process of experimentation, he developed an apparatus in 1909 that used pressure to lower the temperature of the reaction. He also worked out the optimum ratio of hydrogen and nitrogen gases—roughly three to one—to yield the highest ammonia content. However, further improvements to his method were required in order to make mass production viable.

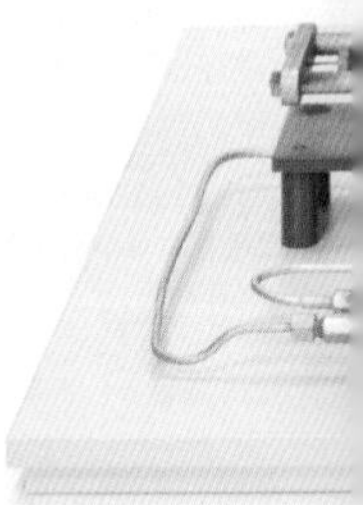

German scientist Haber (1868–1934) won the Nobel Prize in 1918 'for the synthesis of ammonia from its elements'. More controversially, he is known as the 'father of chemical warfare' for his work both to weaponize chlorine gas and other chemical agents and to oversee their deployment in the field.

Haber used this apparatus to demonstrate his process of producing ammonia from hydrogen and nitrogen gases.

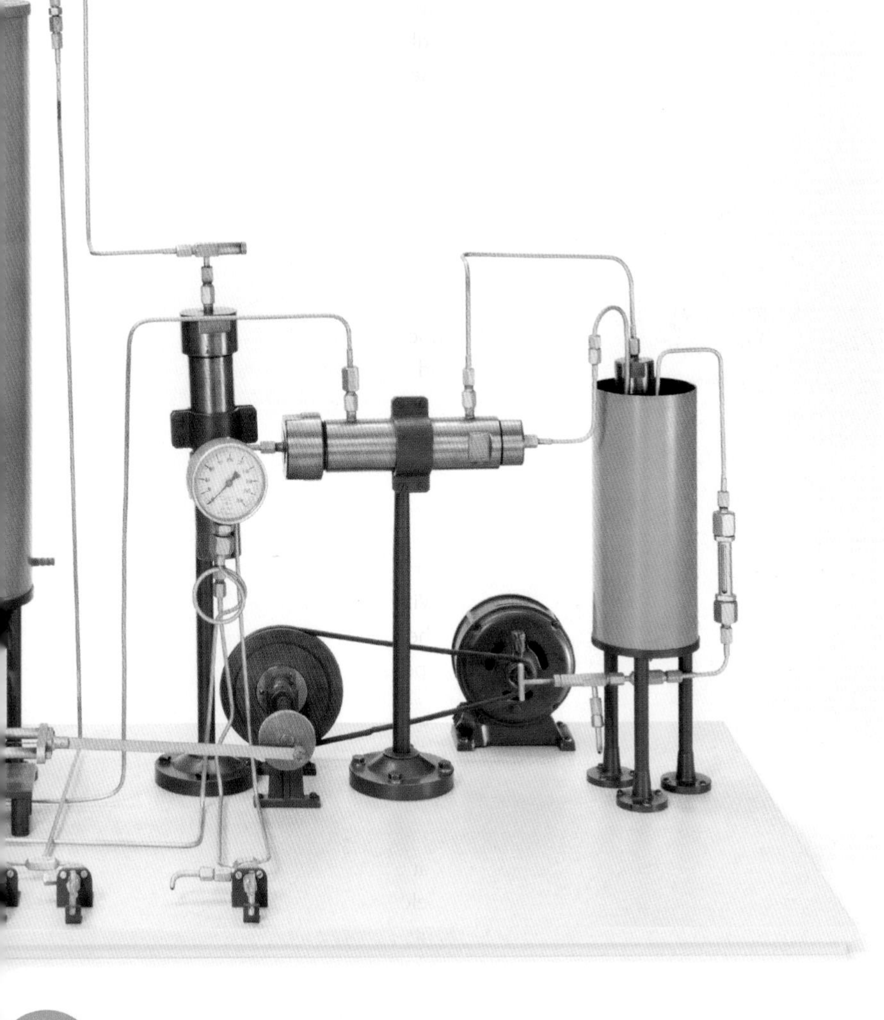

?

Within five years of its invention, the Haber process, as it became known, was refined into a large-scale industrial operation by Carl Bosch at the German chemical company BASF. The firm purchased the process and put it to great use during World War I, not only as a source of fertilizer after the country was blockaded, but also as a means to manufacture explosives. Today, it is estimated that fertilizers originating from the Haber process sustain food production for around half the world's population.

Optical microscopy had been the principal means of examining objects until Wilhelm Röntgen's discovery of X-rays in 1895 gave science a new tool to see into the construction of solid matter. X-rays are smaller than the gaps between each atomic layer of the crystal lattice, so can be used to precisely calculate the distances between atoms in solid matter.

The discovery of X-rays has increased the keenness of our vision 10,000 times; we can now 'see' the individual atoms and molecules.

William Bragg experiments with the ionization spectrometer that he designed and built.

Nobel Prize-winning crystallographer Dorothy Crowfoot Hodgkin cited William's *Concerning the Nature of Things* (1925) as an inspiration in her Nobel Lecture in 1964.

X-RAY CRYSTALLOGRAPHY

LAWRENCE AND WILLIAM BRAGG

1912–13

In the autumn of 1912, young research student Lawrence Bragg examined the X-ray diffraction photographs that had been produced earlier that year by German physicist Max von Laue, who was studying the use of X-rays to determine the arrangement of atoms in crystals. Bragg concluded that because crystals are composed of a regular, lattice-like configuration of repeating parallel planes, he could work out the exact atomic structure of a crystalline object using a simple formula. This is now known in physics as Bragg's law. Bragg shared his theory with his father, William Bragg—Cavendish chair of physics at the University of Leeds, England—prompting the latter to develop the ionization spectrometer. This device would measure an X-ray's diffraction intensity more accurately than in Laue's photographs and would help those wishing to apply Bragg's law. It fired a tightly focused beam of X-rays into the crystal being studied. The refracted beam was then collected and measured in an ionization chamber. Using the ionization spectrometer was more laborious than taking photographs of the diffracted X-rays and the elder Bragg recommended its use only on more complicated compound structures. The Braggs later investigated the structure of sodium chloride, or table salt, and observed it had a chessboard-like lattice structure.

British scientists Lawrence (1890–1971) and William (1862–1942) Bragg were jointly awarded the Nobel Prize in Physics in 1915 for 'services in the analysis of crystal structure by means of X-rays'. The mineral braggite, the first to be discovered through the use of X-ray crystallography, was named after the pair.

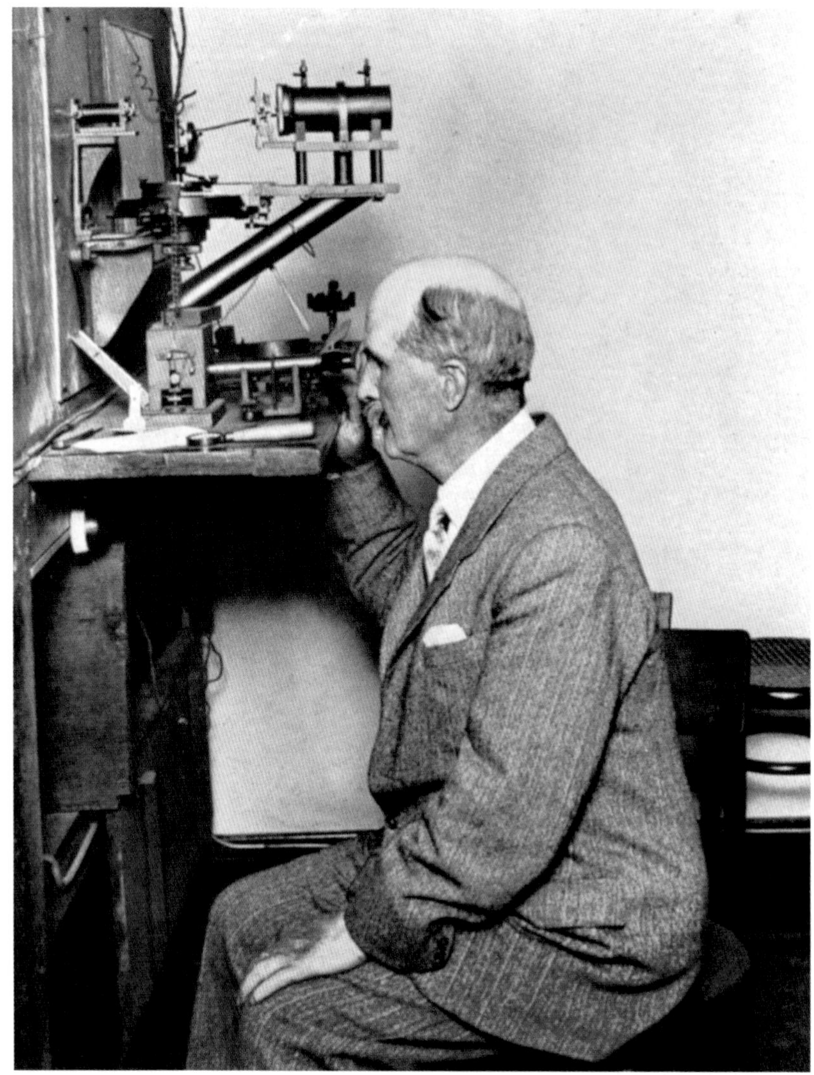

The Braggs' experimental breakthrough with X-ray crystallography gave scientists a means of determining the exact atomic structure of all manner of organic compounds, including the mould penicillin, and even the structure of DNA. Importantly, accurate knowledge of a compound's atomic structure has frequently facilitated its synthesis in the laboratory for mass production.

Barcroft was renowned for his willingness to put his own health on the line in his experiments, and his comparative human and animal studies on the effects of chemical weapons enabled him to measure the consequences of chemical agents in laboratory conditions. His work provided invaluable data on how best to manage exposure and possible recovery times for Allied soldiers exposed to the gas.

EXPOSURE TO CHEMICAL WEAPONS
JOSEPH BARCROFT
c. 1915

British physiologist Joseph Barcroft undertook a series of experiments to establish the toxicity of hydrogen cyanide, a gas that was being used as a chemical weapon. He noted that exposure to the gas, hich inhibits cells' ability to take up oxygen, was not ecessarily lethal in itself and that small doses could be vercome. Consequently, he decided to map to what xtent the density of the gas and the length of exposure ffected the gas's lethality on various animals. He started ith goats, which he placed in chambers in groups of our, and exposed them to different concentrations of ydrogen cyanide gas. Barcroft repeated the experiments ith dogs, cats, guinea pigs and monkeys, before stepping ito the gas chamber himself. It was filled with 500 o 625 parts per million (ppm) of hydrogen cyanide as. In order to complete the comparative analysis, he vas accompanied by a dog. He maintained a similar evel of activity to his companion by following the og's movements. The dog did not fare well, becoming nsteady after fifty seconds and convulsing after ninety econds. At this point, Barcroft felt no ill effects and eft the gas chamber after ninety-one seconds. Several ninutes later, a wave of nausea hit him and after ten ninutes he found it difficult to concentrate. However, oth he and the dog made a full recovery.

The Second Battle of Ypres during World War I marked the first mass deployment of chemical weapons in an offensive capacity. This event led scientists on all sides of the conflict to embark swiftly on studies, away from the battlefield, of the effects of various chemical agents on humans.

Barcroft (1872–1947) served as the chief physiologist at the Gas Warfare Centre at Porton Down, England, during both world wars, and he was one of the foremost experts in the United Kingdom on chemical warfare. Aside from his contributions to this field of study, he is best remembered for his work on haemoglobin and the respiratory function of the blood, for which he was nominated for a Nobel Prize. Barcroft became a fellow of the Royal Society in 1910 and received its Copley Medal in 1943, also for his work on blood and respiration.

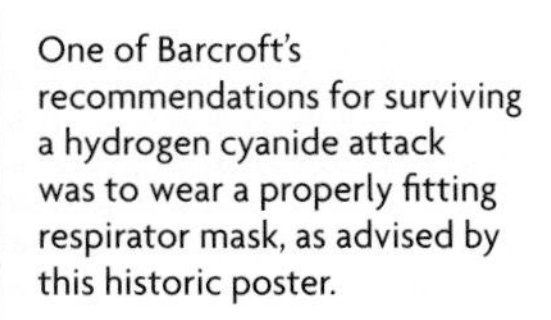

One of Barcroft's recommendations for surviving a hydrogen cyanide attack was to wear a properly fitting respirator mask, as advised by this historic poster.

In a further experiment, Barcroft sealed himself in a glass chamber for six days to assess the effect of a reduced-oxygen environment on the oxygen in the blood.

After Clark had set up the Tuskegee study with Raymond Vonderlehr and later extended it, John Heller (1905–89) inherited the programme in 1943 and allowed its continuation. As director of the venereal disease section of the Public Health Service, he also oversaw the United States' wartime venereal disease control programme and adopted penicillin as a treatment, although this was not made available to the Tuskegee participants. He left the role in 1948 to become director of the National Cancer Institute. In 1961 he was presented with the World Peace Through World Health Award for his work to combat cancer.

The Public Health Service took photographs of its tests throughout the Tuskegee experiment, although the reasons for doing so are unclear.

In 1997 President Bill Clinton formally apologized on behalf of the federal government to the Tuskegee survivors.

SYPHILIS STUDY
TALIAFERRO CLARK
1932–72

The Tuskegee syphilis experiment began as a six-month study on the effects of long-term exposure to syphilis. It was set up by Dr Taliaferro Clark under the auspices of the US Public Health Service. The study involved 600 African American men of low income and between twenty-five and sixty years of age, most of whom were engaged in agricultural work. Many of them were reluctant to participate but they were swayed by talk of regular medical examinations, funeral expenses and even treatment for 'bad blood', Southern slang for a range of conditions that could include syphilis. Of the 600 men recruited, 399 had syphilis and 201 did not. The examinations often included painful spinal taps to test the cerebrospinal fluid for evidence of neurosyphilis, caused by lack of treatment. Doctors elected to continue the experiment, and in 1936 they decided to follow the untreated sufferers until they died. The study progressed under the direction of the Public Health Service until an investigator, Peter Buxtun, raised ethical concerns. He went to the press in 1972, which led to a public furore and the end of the experiment.

In the early 20th century, sexually transmitted diseases were common across swathes of the US population, but treatment was often limited to wealthy and well-educated citizens. Racial prejudice prompted some doctors to consider the merits of studying the effects of long-term exposure to syphilis among poor African American males in Tuskegee, Alabama. They wanted to ascertain whether the disease would operate differently in this study group and how long sufferers could live with the disease.

An official enquiry followed the revelation of the Tuskegee study, but this did not happen until the participants filed a class-action lawsuit against the federal government. This led to a $10 million out of court settlement that included medical treatment for the study subjects, their wives and their children. It prompted a change in US law in the form of the National Research Act (1974) to ensure that future clinical trials were regulated properly and proceeded only with informed consent.

The Tuskegee syphilis study continues to cast its long shadow on the contemporary relationship between African Americans and the biomedical community.

VANESSA GAMBLE, PHYSICIAN

Norwegian biologist and ethnographer Heyerdahl (1914–2002) gained great acclaim for his spectacular expeditions. He was a member of the Norwegian Academy of Sciences and was noted as one of the top five Norwegians of the century in a poll conducted to mark the centenary of the country's independence.

TRANSOCEANIC MIGRATION

THOR HEYERDAHL

1947

Having secured funding to build a small sailed raft, Thor Heyerdahl convinced a crew of five fellow Scandinavians to join him 'sailing' on ocean currents from Peru to Polynesia. He intended to demonstrate that the Inca people of the pre-Columbian era could have colonized the islands in Polynesia by simply drifting on ocean currents. The craft he built was named the *Kon-Tiki* and it was constructed mainly from hemp rope and balsa wood. Supplies for the journey included a mixture of traditional and modern water containers, fruit, nuts and vegetables, as well as US Army rations and radio equipment. Heyerdahl and his crew began the voyage on 28 April 1947 when they were towed outside the shipping lanes off the coast of Peru. They drifted for ninety-seven days, carried by the wind and the Humboldt Current, before sighting land. During this time, they read books and supplemented their supplies with fish. On day 101, the expedition came to an end when the raft hit a reef and was beached on Raroia atoll; the crew were met and rescued by inhabitants of a neighbouring island. Although he had not reached Easter Island, Heyerdahl had reached Polynesia.

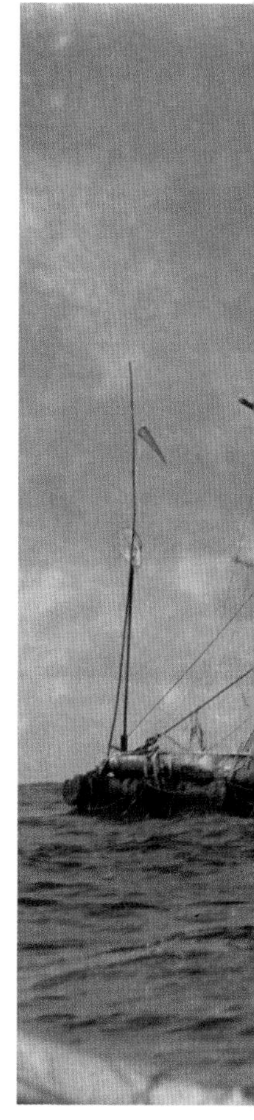

Although the *Kon-Tiki* had sails and rudimentary cross-boards, it had no rudder and was not designed to be steered.

In the 1940s, it was widely believed that Easter Island in the Pacific Ocean had first been colonized by the islands of Polynesia to the west. However, Heyerdahl noted similarities between sculptures on Easter Island and those in pre-Columbian Peru. He set out to prove that South Americans could have crossed the Pacific to reach Easter Island despite the primitive nature of their raft-building technology.

"

The Kon-Tiki *expedition opened my eyes to what the ocean really is. It is a conveyor and not an isolator.*

?

Genetic tests on native inhabitants of Easter Island have shown that they share stronger genetic links with other Polynesians than with modern South Americans. However, Heyerdahl maintained that his expedition was less about whether South Americans actually colonized Easter Island and more about establishing the fact that ancient peoples could have crossed the oceans. He later demonstrated that boats built from papyrus, resembling those seen in ancient Egyptian hieroglyphs, could have crossed the Atlantic.

Despite the many flaws and the fraudulent results of Vicary's cinema experiment, recent investigations have shown that his premise was not entirely without merit. German experimental social psychologists Johan Karremans and Wolfgang Stroebe performed subliminal advertising tests in laboratory conditions in 2006 and concluded that although such messages did not make test subjects want a drink if they were not already thirsty, for those who were thirsty subliminal messages could be used to influence the choice of beverage.

Vicary allegedly inserted this message into the Oscar-winning film *Picnic* (1955), starring Kim Novak, as part of his subliminal advertising experiment.

In the 1940s, the notion of influencing people while they slept piqued the interest of certain psychologists. Lawrence LeShan conducted an experiment to see if he could influence young boys who had a history of biting their nails to stop the habit by playing instructions to do so as they slept. His results were inconclusive.

SUBLIMINAL ADVERTISING

JAMES VICARY

1957

At a press conference held on 12 September 1957, US market researcher James Vicary announced that he had performed an experiment in subliminal advertising over a six-week period at a cinema in Fort Lee, New Jersey. He claimed that he had inserted the messages 'Hungry? Eat popcorn' and 'Drink Coca-Cola' into a film for 1/3,000 of a second every five seconds. According to Vicary, during the period of the experiment, the venue had seen sales of Coca-Cola and popcorn increase by 18.1 per cent and 57.8 per cent, respectively. Although this revelation received a great deal of publicity, Vicary was never able to reproduce the outcomes he had claimed. He later confessed to inflating his results and that his original experiment had produced only 'a small amount of data—too small to be meaningful'. However, by this point, the experiment had seized the popular imagination.

Vicary (1915–77) made a significant impact on both science and culture, and his work prompted investigations by the CIA and broadcast regulators into the potential applications for subliminal advertising, even though there was little evidence of its effectiveness.

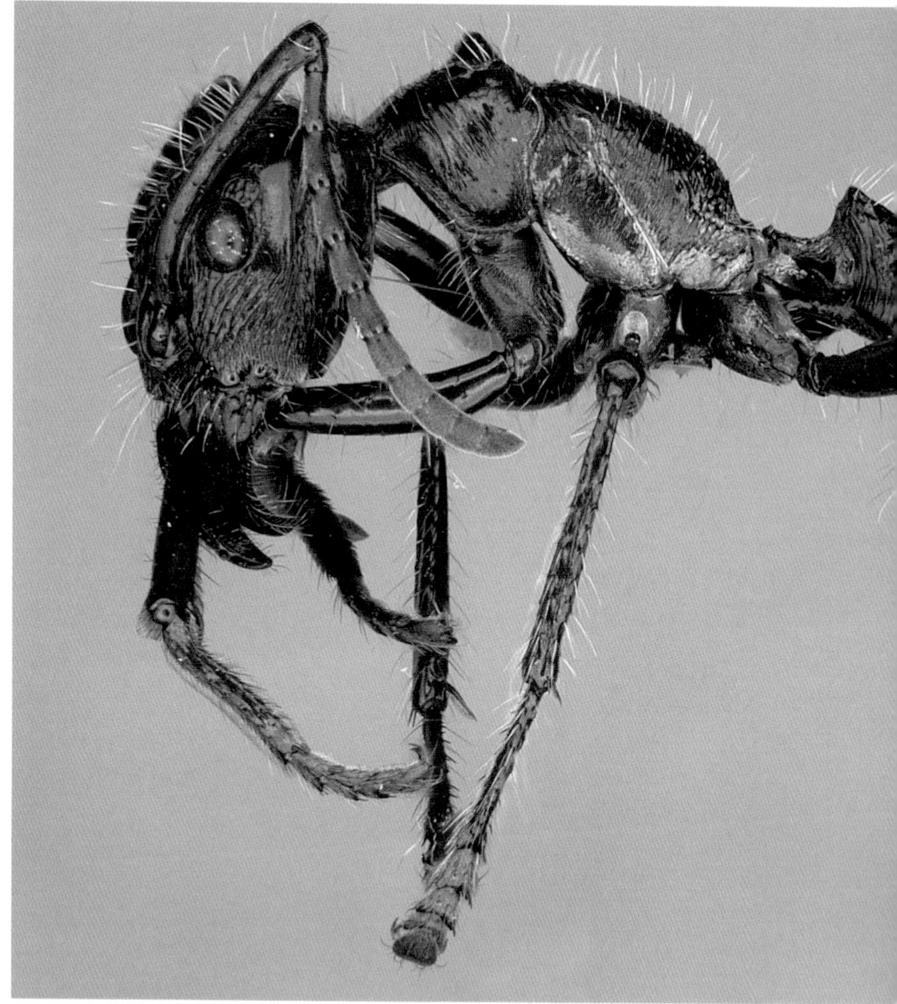

Various scales to measure the pain caused by insect stings have been proposed by different entomologists over the years, but Schmidt's vivid and detailed descriptions of the sensation of each sting captured the popular imagination. Despite the simple, four-point index listing only seventy-eight species, its basis in first-hand empirical experience gives it a solid scientific grounding.

The bullet ant (*Paraponera clavata*) has a large, retractable, syringe-like sting at the end of its abdomen.

During the process of evolution, many insects have developed venom delivery mechanisms to protect themselves and their colonies. By the early 1980s, insect anatomy and defence behaviour had been studied widely, but the comparative experience of pain that venom caused in humans was less well-defined.

SCHMIDT PAIN INDEX

JUSTIN SCHMIDT

1983

For US entomologist Justin Schmidt, insect stings were a common hazard that he faced while collecting specimens for study. Although he did not set out to get stung, he put his experiences to good use by devising a pain index to systematically record and rate the agony of the stings he suffered. His index rated the pain of each one on a scale of 1 to 4, but within that hierarchy he also described the sensation produced by each species. The sweat bee was rated 1, with its sting likened to 'a tiny spark [that] has singed a single hair on your arm', whereas the tarantula hawk was rated 4, for a sting that is 'blinding, fierce [and] shockingly electric', like 'a running hairdryer has been dropped into your bubble bath'. The most painful sting was that of the bullet ant, rated 4+ and described as 'pure, intense, brilliant pain. Like walking over flaming charcoal with a 3-inch nail in your heel'.

The Schmidt pain index was referenced in the movie *Ant-Man* (2015), with the title character harnessing the 'super' power of bullet ants to aid him in his escapades.

Schmidt (1947–) is director of research at the Southwest Biological Science Center, in Flagstaff, Arizona, where he studies the chemical and behavioural defences of ants, wasps and arachnids. In 2008 he announced plans to update his index to include the insect stings he had suffered since 1990.

The availability of Coca-Cola and its history as a patent medicine—one that claims curative powers that it does not possess—continue to make the beverage a popular subject of study. Some claims assign benefits to the drink that are not borne out under scientific scrutiny, but others show results that warrant further investigation. Meanwhile, the brand's ongoing success around the world provides grist to psychologists and social scientists seeking to fathom out the secret of its popularity. There remains much scope for the further scientific study of Coca-Cola.

The exact formula of Coca-Cola is a trade secret. Over the years scientists have analysed the drink to ensure it does not contain banned chemicals.

TESTING COCA-COLA DOUCHES

DEBORAH ANDERSON ET AL.

1985

In the 1980s, a team of US scientists, including gynaecologist Deborah Anderson, tested the claims that Coca-Cola could be used as an effective spermicide. They poured samples of different colas into test ubes containing human semen at a ratio of five to one. nitial results indicated that Diet Coke killed all the sperm : was exposed to, whereas other colas destroyed only round 60 per cent—a poor result for a spermicide. nderson warned that her results should not be viewed s evidence that Diet Coke is an effective spermicide ecause test tubes are not analogous to vaginas. She noted sperm] can make it into the cervical canal, out of reach f any douching solution, in seconds'—faster than anyone an apply the 'shake and shoot applicator', as the Coke ottle became known. Follow-up studies led by Chuang- e Hong showed that none of the Coke samples performed etter than 70 per cent at killing sperm. He concluded nat any spermicidal effect present was so weak as to nake the beverage inadvisable as a contraceptive method. nce that time, scientists have explored another use for ne drink: as a treatment for phytobezoars, which are lumps f undigested fruit and vegetables that become trapped the stomachs of patients suffering from gastroparesis.

Some of the more promising results have been found by Greek scientist Spiros Ladas. Reviewing the literature from 2002 to 2012, concerning Coca-Cola as a treatment for phytobezoars in forty-six patients, he learned that Coke had been used as part of an effective treatment to move the masticated masses in 91.3 per cent of cases, and it had proven effective on its own for half of the patients cited.

Taiwanese doctor and researcher Hong (1949–) has received a number of honours throughout his long career, including awards from the Taiwan Physicians' Association and the British Pharmacological Society. In addition to his medical research work, which resulted in several patents, Hong's observations netted him a share of the Ig Nobel Prize in Medicine in 2008, shared between his own team of investigators and that of Anderson, for seemingly contradictory evidence on the spermicidal properties of Coca-Cola. Hong now serves as director of Taipei Medical University and in 2015 published a biography titled *Encounters with Medicine and Sperm*.

Coca-Cola douches had become a part of contraceptive folklore during the 1950s and 1960s, when other birth control methods were hard to come by.

In addition to his research, which includes the creation of a cultured neural network to remotely control a simple robot, Warwick (1954–) has actively promoted science to the young. He presented the Royal Institution Christmas Lectures in 2000, a annual series of scientific talks for a non-specialist audience.

HUMAN CYBORG
KEVIN WARWICK
1998

On 24 August 1998, British engineer Kevin Warwick became the world's first human cyborg when a small silicon chip transponder encased in glass was surgically implanted into his forearm. Employing the same technology that is now used in contactless payment systems, the experiment allowed a computer to track Warwick's movements and to respond to his presence as he walked around the department of cybernetics at the University of Reading. With automated systems in place to identify the unique signal emitted by his implant, he could operate doors, turn on lights and gain access to computers without lifting a finger. The second phase of the experiment occurred in March 2002, when doctors implanted a different silicon chip, known as the one hundred electrode array, into the median nerve fibres of Warwick's left wrist. This neural interface enabled him to control a robot arm with his own neural impulses, even at a great distance. Using the same device, and a similar one implanted into the wrist of his wife, Irena, Warwick was also able to demonstrate bidirectional neural interface, with sensation from Irena's hand being felt by Warwick, and vice versa.

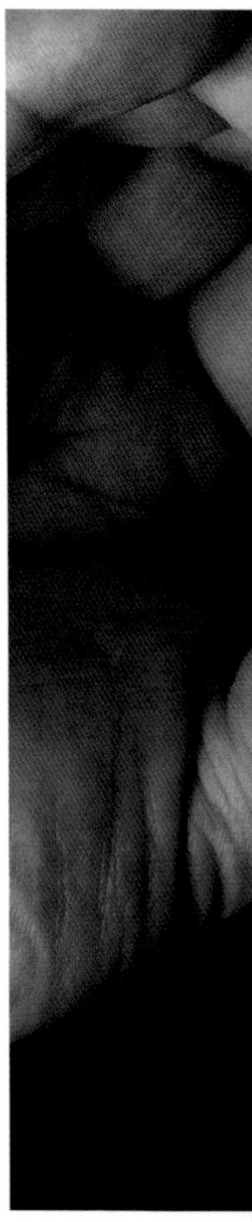

Warwick holds a transponder similar to the one implanted in his arm. It made him the first human cyborg.

The term 'cyborg' originates in the article 'Cyborgs and Space' (1960) by Manfred Clynes and Nathan Kline, but the idea of artificial enhancements that give humans extra abilities can arguably be traced back to the first tool-using hominids. However, it is only in the 21st century that technologically advanced, surgically implanted augmentations have become more widely available.

I was born human. But this was an accident of fate—a condition merely of time and place. I believe it's something we have the power to change.

?

Although Warwick has been derided as a self-important 'Captain Cyborg', his work has had practical applications. Transponders similar to the ones implanted in him have valuable medical applications because they can communicate the identity and medical needs of patients whose conditions may prevent them from informing doctors themselves. In addition, bidirectional neural interfaces raise the possibility of extending the limits of the human body, with opportunities for new mechanical inputs and outputs.

US molecular biologist Lewis has contributed a significant amount of research to the formation of spider silk, helping to isolate the gene responsible for the spider silk protein in 1990. He licensed the technology to Nexia.

The orb spider's dragline silk, if made as thick as your thumb, can support a fully loaded jet.

JEFFREY TURNER, NEXIA FOUNDER

A great deal of the potential for transgenic plants and animals has not yet been fully realized, and this is true of spider-goat silk. Possible uses run to everything from organic sutures for ligaments and tendons in the medical industry to lightweight bullet-proof clothing. However, before any of these applications can reach the market, production of the silk will need to be raised substantially and, particularly in the case of medical applications, the product will need to be thoroughly tested and approved.

In 1999 scientists at the Canadian biotech firm Nexia successfully cloned goats, utilizing techniques similar to those used to create Dolly the sheep in 1996. However, goat cloning was not the end goal of this experiment. It was merely the first step to creating transgenic animals: creatures that contain genetic material and traits of two different species. The aim was to insert the genes for creating the dragline form of spider silk, which is prized for it high tensile strength, from the golden orb-weaver (*Nephila clavipes*), also known as the golden silk spider, into goat embryos. Eventually, the scientists succeeded in creating 'spider-goats': goats that expressed the protein for spider silk in their milk. In 2002 the company's chief executive referred to this goat milk silk as the world's first transgenic material, but unfortunately it did not save Nexia from bankruptcy in 2009. Before it went out of business, the firm trademarked the spider-goat silk as BioSteel, to indicate the prospective product's strength. Although some of the spider-goats were sold to the Canada Agriculture Museum, one of the scientists, Randy Lewis, continued to work on creating a transgenic source of spider silk, and in 2010 he bred three goats capable of expressing spider silk milk. Once collected, the milk needs to be processed before the silk can be harvested and spun into a usable fibre. Lewis has also worked on adding the *N. clavipes* silk gene to *Escherichia coli* bacteria and even to the alfalfa plant, and each new transgenic creation is assessed for its ability to increase the spider silk yield.

These transgenic goats raised on a farm at Utah State University produce milk with high levels of antimicrobial protein.

Orb-weavers are not suitable for conventional farming because of their tendency to eat nearby rivals.

Selective breeding has been a part of farming for millennia, but synthetic genetics takes selective breeding to the next level. The target traits do not need to come from the same animals; the resultant offspring have characteristics from two completely different species.

Proteins are the tools that the body uses to maintain and repair itself. They work by changing shape, or folding. The process of folding takes nanoseconds and it is not always successful; misfolding can lead to serious health consequences.

FOLDING@HOME
VIJAY PANDE
2000–present

Studying the folding of proteins can help doctors to better understand disease and therefore to develop new treatments. With this in mind, US biomedical scientist Vijay Pande devised a piece of software called Folding@home. It was designed to take the vast amount of calculations that is involved in simulating the folding of a protein and break it into smaller chunks. These are then distributed to any volunteers who are willing to download the software to their personal devices and allow the calculations to be made while they are not using the device. The resultant simulation data is fed back to experts, who examine it for potentially significant results. The first Folding@home software was released for home users on 1 October 2000, and since that time more than eight million people have downloaded it to their computers, tablets, mobile phones and even games consoles. With enough clients linked together, Folding@home effectively becomes a huge supercomputer.

Distributed computing is a natural and powerful way to do research. When we started Folding@home, there were no other [distributed computing] projects studying biology or chemistry or physics.

One of the world's top innovators and a professor at Stanford University, Pande (1970–) received the Thomas Kuhn Paradigm Shift Award in 2008 for presenting 'original views that are at odds with mainstream scientific understanding'.

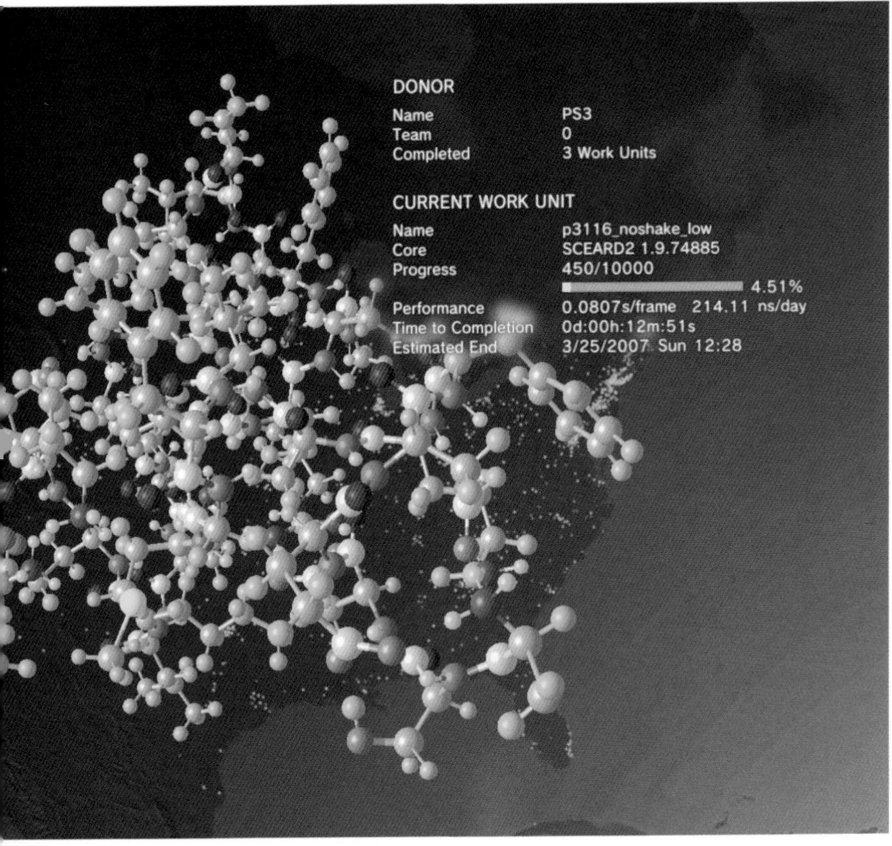

Since 2006 Folding@home has had a particular focus on the consequences of protein misfolding, which plays a prominent role in the development of Huntington's disease, Alzheimer's disease and a number of other high-profile ailments. Data from the project has also been significant in the development of innovative treatments that rely not only on new medicines that have not yet been approved by regulators, but also on existing ones that have been repurposed for different therapeutic effects. By the beginning of 2016, the Folding@home distributed supercomputer had an output of more than 16,000 teraflops of computing power and 129 scientific papers had been published citing data from it.

The Folding@home client for PS3 allows gamers to contribute their computing power to scientific research.

Non-violent deterrents have become increasingly important for security and policing in recent years, and, like Stapleton, a number of scientists and engineers have developed sonic devices. One notable example is the Long Range Acoustic Device, or LRAD. It not only works as an effective loud hailer, but it is capable of producing a 2.5 kHz beam of sound that can be heard many miles away. At best this is unpleasant, but it can cause hearing loss if turned up to 160 dB. Fortunately, this device is not widely available to the public.

Stapleton displays his Mosquito anti-loitering device, installed at the test site—the Spar convenience store—in Barry, Wales.

ANTI-LOITERING DEVICE
HOWARD STAPLETON
2005

British electronics engineer and inventor Howard Stapleton began to experiment with equipment to generate ultrasonic frequencies in 2005. He was aware that humans slowly lose the ability to detect high-frequency sounds as they get older. Such loss is known as presbycusis and it affects higher frequencies at a roughly regular rate. This knowledge gave Stapleton a starting point in his search for the correct frequencies to use for an alarm that would affect only teenagers. As his target frequencies lay outside of his own hearing range, he resorted to testing the device on his children, eventually settling on 16.8 kHz modulated at 4 Hz, a frequency that is audible almost exclusively to teenagers and children. In order to make the sound more unpleasant, Stapleton opted to modulate the tone. Field trials followed at a local convenience store, whose manager had complained of troublesome teenagers. With his device placed above the store entrance, the inventor hid in his car and activated the alarm when youths began to congregate by the door. The noise soon drove them away. Stapleton followed up this experiment with an extended trial in which the store owner operated the alarm—dubbed the Mosquito—and achieved similar results.

High-frequency ultrasonic welding was pioneered in the United States by Robert Soloff and Seymour Linsley in the 1960s. Stapleton first encountered high-frequency sound welding at a factory, where he found the noise markedly more unpleasant than the older employees. Years later, when he was looking for ways in which to deter loitering teenagers, he decided to explore unpleasant high-frequency sounds further.

Stapleton (1966–) began his career as an electronics apprentice at British Aerospace and worked in hotels before coming to the security industry and developing the Mosquito anti-loitering device. Stapleton's unusual alarm won the Ig Nobel Prize for Peace in 2006, awarded for 'inventing an electromechanical teenager repellant—a device that makes annoying high-pitched noise designed to be audible to teenagers but not to adults; and for later using that same technology to make telephone ringtones that are audible to teenagers but probably not to their teachers'. He continues to develop security-related products for his company Compound Security.

I didn't want to make it hurt, it just has to nag at them. . . . It's very difficult to shoplift when you have your fingers in your ears.

DINOSAUR GAIT
BRUNO GROSSI ET AL.
2014

In an experiment to simulate how theropod dinosaurs—the group of meat eaters that includes the tyrannosaurs—might have walked, Bruno Grossi and his colleagues took twelve two-day-old domestic chickens and divided them into three groups. Chickens in the control group received no modification, those in the second group had a clay weight added to their posteriors and those in the test group were fitted with a clay weight that had a long wooden stick projecting from it. The additions were 'adjusted to the shape of each chicken's pelvic girdle, making the stick continuous with the projection' in the manner of a dinosaur's tail. The weights were restricted to around 15 per cent of the chicken's weight and they were attached to an elastic bodysuit worn by the bird. The suits and weighted tails were replaced every five days as the chickens grew, but otherwise they remained on the animals all the time to ensure that the birds adapted to the change in the centre of gravity that the tails provided. When the chickens were twelve weeks old, Grossi started to film them walking along a 3-metre (10 ft) track. The scientists plucked the feathers from around the tail region and marked joints with tape to help trace the chickens' movements more precisely. They soon observed that the test group had a more vertically orientated femur when standing and that the bone had a range of motion around three times greater than that of the chickens in the control group.

The extra weight and length of the artificial tails affected the chickens' centre of gravity and their stance, as well as how they walked.

Dinosaurs roamed Earth for millions of years until they went extinct at the end of the Cretaceous period some 66 million years ago. A vast amount of information has been gathered from their fossilized remains, but there are many areas of their behaviour, such as how they walked, that remain uncertain, leaving biologists to look to dinosaurs' closest living relatives, birds, for experimental evidence.

Although the test chickens were treated well while they were alive, they were killed as part of the experiment so that their bones could be analysed.

Chilean biologist Grossi (1978–) and his team shared the 2015 Ig Nobel Prize in Biology for their experimental simulation of how dinosaurs might have walked. Grossi proved that he was willing to put himself through similar tests by attaching a plunger to his posterior to demonstrate his own dinosaur gait.

[We expect to provide] a more nuanced understanding of the relationship between form and function in dinosaur evolution.

With no Cretaceous dinosaurs available for study, it is impossible to judge the accuracy of Grossi's modelling. However, by observing the test group throughout its growth and ensuring that the chickens had their weighted tails attached at all times, he produced a more thorough study than in previous chicken/dinosaur research. The team was also keen to note the limits of the experiment, particularly in the light of recent research showing that reduced forelimbs, in addition to a longer tail, affected the stance of dinosaurs. This is difficult to emulate on birds without damaging them, but the team hopes its work will open up new avenues for dinosaur locomotive research.

CHAPTER FOUR
THE PLANET

The planet on which we live and its associated atmosphere and oceans have long been an important constant in scientific enquiries. The depths of its oceans and the dynamic processes that shape it remained unknown until they were elucidated by experiments conducted surprisingly recently. Meanwhile, Earth's composition has been broken down into various elements, classified in the periodic table by Dmitri Mendeleev, and isolated experimentally by scientists, including Joseph Priestley (oxygen) and Marie Curie (radium and polonium). Furthermore, the uniform radioactivity of some elements has led to the formation of experiments by which scientists can not only date organic materials produced on Earth but also the planet itself.

‹ Underwater Living (see p. 172)

Lightning rods became a popular safety measure to protect buildings from the dangers of lightning strikes. However, controversy developed as to whether the ends of the rods should be pointed, as Franklin advocated, or rounded, on the grounds that pointed rods might increase the risk of lightning strikes. Experiments were conducted in the late 18th century to test both designs, but the results were inconclusive.

LIGHTNING ROD
BENJAMIN FRANKLIN
1750

Among the many experiments that Benjamin Franklin is thought to have conducted with electricity, the best known involved him flying a kite during a storm in 1752. He was ntending to draw down electricity into a key and collect t in a Leyden jar, thus demonstrating the electrical harge of lightning. Although there is some doubt as to hether he could have conducted such an experiment the way he described without killing himself, he erformed an electrical experiment two years earlier that vas perhaps more consequential. Franklin had noticed ow a sharply pointed iron needle could be used to raw what he described as 'electrical fire' from an object harged with static electricity. The electricity seemed to e attracted to the metal. He also noted that if he was olding the needle, and therefore grounding it, more lectricity seemed to be drawn out. This led the scientist o propose attaching a long iron rod, which extended at east 60 centimetres (2 ft) into the ground, to the side f a tall building so it could similarly draw down the lectricity from storm clouds and quickly drain it into the round. In turn, this would reduce the risk of lightning itting and damaging a building's structure. Franklin nstalled such a device on his own home, and over time rchitects around the world took note and followed suit.

A complete physical theory of how lightning works is still hotly contested today, but by the mid 18th century, most scientists agreed that it could be highly dangerous, especially to buildings and other tall man-made structures. The question was whether anything could be done to mitigate the risk of lightning strikes.

Franklin (1706–90) was something of a renaissance man: he was a writer, printer and publisher, politician and ambassador. He was also an experimental scientist and inventor. The Franklin Institute, a museum and research centre dedicated to science, was established under his name in his native Philadelphia, Pennsylvania, in 1824, and it presents the prestigious Benjamin Franklin Medals in various categories for achievements in science and engineering.

An iron rod . . . will receive the lightning at its upper end, attracting it so as to prevent its striking any other part [of the building].

Franklin's electricity experiments and invention of the lightning rod gave him world renown. Here, he is depicted on a promotional card produced by a French meat extract producer.

The credit for the discovery of oxygen is highly contested, but the honours for the isolation of the gas rest with Priestley. However, French chemist Antoine Lavoisier was responsible for correctly interpreting that Priestley's results demonstrated that oxygen is the gas that is consumed in combustion reactions. He also named the gas 'oxygen', meaning 'acid creator', based on the observation that its combustion products were often acidic. His alternative to phlogiston theory was widely accepted and helped to make chemistry a systematized discipline.

Lavoisier's large-scale experiments, such as the combustion one depicted here in 1874, commanded attention, but it was his naming of elements that provided a lasting legacy.

DISCOVERY OF OXYGEN

JOSEPH PRIESTLEY

1772

Experimental chemist Joseph Priestley first produced a gas that he termed 'dephlogisticated air' in 1772. It was the result of a series of experiments with various gases that he viewed as simply different kinds of air. His most significant experiment involved heating the chemical compound known as red mercury calx (red mercury oxide) and capturing its gaseous discharge into sealed glass containers. When Priestley tested the qualities of this 'superior air' by placing a lit candle in one of the sealed containers, he was surprised to discover that the flame burned longer and more vigorously than it did in a similar container of 'ordinary air'. He also observed that red-hot wood crackled and returned to flame in the presence of the dephlogisticated air. Priestley then tested this gas on animals. He placed two mice into two separate sealed glass containers, one of which was filled with ordinary air and the other with his newly discovered gas. The mouse in the ordinary air asphyxiated after about fifteen minutes, whereas the animal in the gas survived for more than an hour and was taken out alive and 'vigorous'. Priestley later repeated the experiment with the mouse in dephlogisticated air for the Royal Society.

Rivals Priestley (1733–1804) and Lavoisier (1743–94) are widely considered to be the parents of modern chemistry, and many of their individual contributions rely equally on those made by their peer.

In the mid 1750s, many early chemists believed in phlogiston theory: the idea that there was a noxious gas—phlogiston—contained in various combustible objects and that this was released during the process of burning. The fact that candles are extinguished and that animals die in sealed spaces was seen as evidence that only a finite amount of phlogiston could be absorbed by air.

After the accuracy and purpose of his periodic table was recognized, Mendeleev (1834–1907) was appointed director of the bureau of weights and measures in St Petersburg in 1893, a position that he held until his death. Perhaps as a result of this, he has been widely credited for the adoption of the metric system of measurement in Russia. He was awarded the Royal Society's Davy Medal in Chemistry in 1882 and its Copley Medal in 1905. However, the most fitting tribute to Mendeleev's work is the fact that the element mendelevium bears his name.

This version of the periodic table was handwritten by Mendeleev in 1869. It predates the more diagrammatic version that is commonly used in chemistry lessons, with all its columns and groups.

Mendeleev's table was remarkable because it predated the discovery of the atom and the atomic number, the underlying cause of each element's properties.

PERIODIC TABLE
DMITRI MENDELEEV
1869

As a teacher, Dmitri Mendeleev studied the various properties of all the known elements and also kept abreast of new elements as they were discovered. He arrived at the particular arrangement of his periodic table not through the analysis of his own experiments, but as a result of the careful examination and critique of those of his peers. His grouping was not based solely on the single criteria of atomic weight; it also cross-referenced similarities in each element's properties. However, his smartest move was that he did not assume that every element had already been discovered. He left gaps in the table between the known elements where he predicted undiscovered elements should exist, based on the weights and properties of those closely related. Mendeleev even systematically named these predicted elements, such as eka-aluminium which was situated one below (eka-) aluminium on his table. In 1875 French chemist Paul-Emile Lecoq de Boisbaudran discovered what was later named gallium, which he obtained through electrolysis in small amounts from the ore zinc blende. Subsequent analysis showed that gallium matched Mendeleev's prediction for eka-aluminium. However, the chemist was not infallible: he failed to predict the inert noble gases, but this was an understandable omission due to their lack of reactivity.

By the mid 19th century, scientists had isolated some sixty elements, but there was little agreement on how they related to each other. German chemist Johann Döbereiner and British chemist John Newlands attempted to group them, but their systems lacked the simplicity and accuracy of Mendeleev's table.

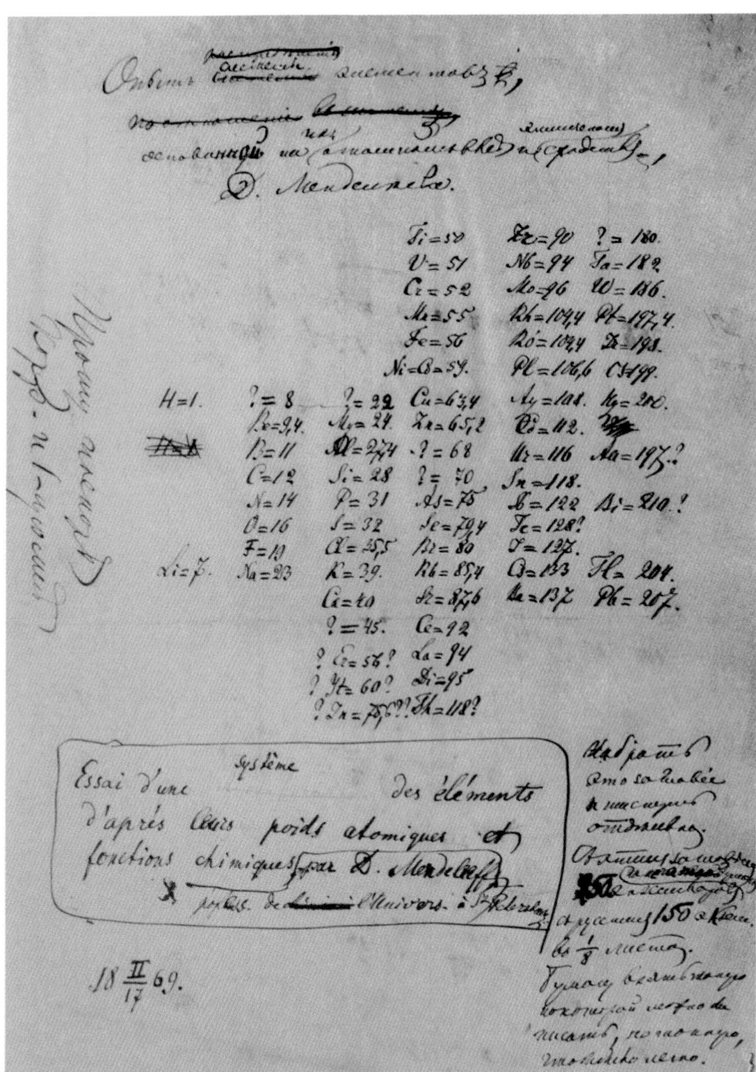

Essai d'une système des éléments d'après leurs poids atomiques et fonctions chimiques par D. Mendeleeff

18 II/17 69.

Mendeleev's periodic table not only created an elegant method of summarizing data on all elements, it also gave structure to chemistry, revolutionizing how the subject was taught and how chemists viewed their discipline. It helped chemists to predict the properties of synthetic elements and to surmise at a glance how various elements might interact based on their relative positions in the table.

The late 19th century was an exciting time for important scientific discovery. A flood of new phenomena—from X-rays to alpha, beta and gamma rays—was being uncovered, but the mechanisms through which they were produced and the rules that governed their behaviour were not well understood.

DISCOVERY OF POLONIUM AND RADIUM

MARIE CURIE

1898

Many late 19th-century scientists focused their research on X-rays, but Polish-born French physicist Marie Curie decided to study 'uranic rays', the yet to be explained process of radioactivity discovered by Henri Becquerel in 1896. In her initial experiments, she used a sensitive electrometer—designed by her husband, Pierre, and brother-in-law, Jacques—to measure the tiny electrical current generated by the uranic rays emanating from uranium salts. She then used the same equipment to measure similar emissions, which she termed 'radioactivity', from a variety of different ores and elements. During her research, she observed that the rays emanating from the uranium ore pitchblende (now known as uraninite) were much stronger than those from uranium salts. Marie surmised that pitchblende must contain an extra component that was responsible for these stronger rays. At this point, Pierre put aside his own research to work with his wife on the laborious process of separating out various constituent parts within pitchblende in search of the mysterious, highly radioactive fraction that they were detecting. The couple identified not one, but two new elements: polonium and radium.

Marie (1867–1934) is the only woman to have won two Nobel prizes in two separate categories: Physics and Chemistry. Numerous institutions and museums have been named after her, as has the element curium and a unit for measuring radioactivity. She even has three minerals named after her: curite, sklodowskite and cuprosklodowskite (the last two refer to Curie's maiden name).

Marie's new laboratory was much better equipped than the 'miserable old shed' in which she discovered polonium and radium.

It was really a lovely sight . . . the glowing tubes looked like faint, fairy lights.

The discovery of polonium and radium challenged some of the established thinking in physics and chemistry. It opened new areas of scientific research and heralded a potential new treatment for cancer. In addition, the Curies' unwillingness to patent their work and methods facilitated the development of new industrial applications for radioactive materials.

Rutherford's experiment demonstrated that the atom's positive charge must be densely packed into a tiny central nucleus surrounded by a sparse cloud of negative charge, and not evenly distributed as previously thought. This new atomic model led to a fresh understanding of how matter was composed, thereby opening the door to the study of subatomic particles and a whole new domain of particle physics.

ANATOMY OF THE ATOM

ERNEST RUTHERFORD

1909–13

British physicist Ernest Rutherford conducted a series of experiments that resulted in a new understanding of the structure of the atom. They are often known as the Geiger-Marsden experiments after his sistants, Hans Geiger and Ernest Marsden, who performed uch of the hands-on work. In all of the experiments, a urce of ionizing alpha particles—high velocity helium iclei produced during radioactive decay—was targeted a thin metal foil (gold foil was used in the final periment.) A movable fluorescent screen was employed track the path of the alpha particles and to detect iether they were being deflected by the foil. The ientists expected the alpha particles to penetrate the il and to continue in an unimpeded straight line, but ey were, in fact, deflected at a variety of angles, metimes back in the direction of their source. 'It was nost as incredible as if you fired a 15-inch shell at a ece of tissue paper and it came back and hit you,' noted itherford. After refining the experiment, he concluded at the mass of an atom must have a concentrated and ghly charged centre in order for it to repel the alpha rticles in such a way. His calculations were later proven specially redesigned versions of the gold foil periment, conducted by Geiger and Marsden in 1913.

In the late 19th century, atoms were thought to be the smallest indivisible parts of matter. However, with the discovery of radiation in the earliest years of the 20th century, scientists began to uncover evidence of the existence of smaller subatomic particles. Competing theories appeared, but establishing the structure of the atom required scientific proof.

Widely credited as one of the greatest experimental scientists of the 20th century, Rutherford (1871–1937) was the first research student at the Cavendish Laboratory of the University of Cambridge, where he met J. J. Thomson. He won the Nobel Prize in Chemistry in 1908 for his investigations into the disintegration of the elements and the chemistry of radioactive substances. He was later credited with discovering how to split an atom, although he never actually performed the experiment himself. In 1997 the element rutherfordium was named after him.

Rutherford (right) at the Cavendish Laboratory in Cambridge, where he became the facility's director in 1919.

The gold foil experiment arose when Geiger and Rutherford observed an anomaly in the deflection of alpha particles in an early version of Geiger's counter that measures ionizing radiation.

Holmes (1890–1965) is hailed as the father of modern geochronology. He was an early champion of Alfred Wegener's theory of continental drift and theorized the existence of convection currents and movement within the mostly solid rock of Earth's mantle. In 1956 he won both the Wollaston and Penrose Medals from the Geological Society of London and the Geological Society of America, respectively. A crater on Mars is named after him.

[This may] help the geologist in his most difficult task . . . unravelling the mystery of the oldest rocks of the Earth's crust.

The process of dating the age of Earth was complicated by the requirement of finding rocks old enough. As the continents drift further away from each other forming new rock, depicted here, the older rocks are reclaimed by subduction as they sink back into the earth.

RADIOMETRIC DATING

ARTHUR HOLMES

1910

In 1910 British geologist Arthur Holmes began to explore a practical method for dating rocks. It rested on the notion that over billions of yea lead is produced by the slow radioactive decay c uranium. In order to test his method, Holmes sought samples of the oldest rocks that had not been subjected to volcanic action, the age of which had been established. The rocks he used came from deposits in Norway and they were thought to be from the Devonian period. They varied in mineral conte and Holmes employed a number of techniques to ascertain the ratio of lead to uranium for each mineral. He then compared his results to those of Bertram Boltwood, a US physicist who had been the first to establish that lead was the final decomposition product of uranium. Boltwood had performed similar experimen on deposits from a variety of geological periods and locations across the United States, and he had noted a correlation between older rocks and high ratios of lead to uranium. From their combined data, Holmes assembl a rudimentary calibration that gave his original Norwegi sample an age of 370 million years, well within the Devonian period. It was enough to confirm that radioacti decay could be used to measure the age of Earth. Today this technique is known as radiometric dating.

At the beginning of the 20th century, there was little agreement on the age of Earth. The fossil record only captures the remains of animals with bones, shells and cartilage, and it covers only the past 541 million years in any great depth. Earlier rocks exist, so geologists needed to find a better way to date them and to ascertain an accurate age of Earth.

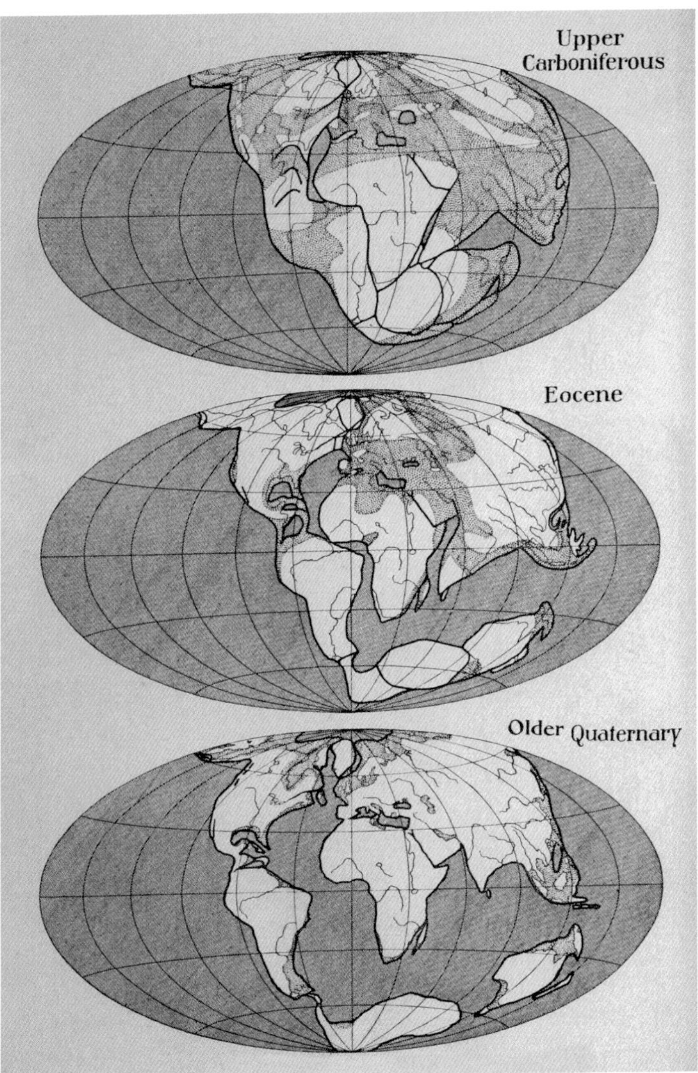

Based on his radiometric dating, in 1927 Holmes estimated the age of Earth to be between 1.6 and 3.0 billion years. In 1956 US geochemist Clair Cameron Patterson employed similar radiometric dating methods, but instead of dating rocks from Earth directly, he used meteorites. His results indicated an age close to the current estimate for the age of Earth: around 4.5 billion years old.

Although the pitch drop experiment may seem trivial at first glance, with only nine drops recorded to date, it has posed scientists some interesting questions over the years, including how they might calculate the average viscosity of the pitch sample involved. The experiment also teaches scientists and common observers alike far more about the scale of science. It underlines the fact that experiments can take decades to reach fruition—sometimes outliving their instigators—and that their impact can reverberate through generations.

Despite being the custodian of the experiment for many years, John Mainstone, seen here with the apparatus, has never witnessed a pitch drop.

PITCH DROP
THOMAS PARNELL
1927–present

In 1927 British physicist Thomas Parnell began preparations for what would prove to be the longest running experiment in history. He intended to demonstrate to his students the surprising qualities of tar itch. Having obtained a small sample of the resin, he egan heating it until it was fluid enough to pour into glass funnel with a sealed stem. Parnell then left the itch to settle for three years. In 1930 he returned to he experiment and placed the funnel in a purpose-built ietal stand, opened the bottom of the funnel and put collecting jar beneath it. He then placed the whole xperiment underneath a large bell jar. The pitch was ow free to flow, and over a number of years a drop of itch slowly formed. As the funnel was not kept in a ontrolled environment, the flow rate of the pitch varied ith the seasons. In subsequent years, Parnell and those ho followed him kept a record of the year and month which each drop finally detached itself and landed the collecting jar. Only two drops fell during Parnell's fetime. The first, and so far only, scientific paper on he experiment, published in 1984—after six drops had llen—estimated the viscosity of the pitch to be about 0 billion times greater than that of water.

After graduating from St John's College, Cambridge, Parnell (1881–1948) moved to Australia in 1904 to become the first professor of physics at the University of Melbourne. In 1911 he moved to the University of Queensland, under whose auspices he published a paper on a high-precision method of measuring inductance. Although his most famous experiment outlived him, this has not prevented his name from being attached to the paper that explores the results. Parnell was also posthumously awarded the Ig Nobel Prize in Physics in 2005 in recognition of his work on the viscosity of pitch. He shared the honour with the experiment's then-custodian, John Mainstone.

In the early 20th century, tar pitch was not a household commodity, but it was reasonably easy to source and widely used in waterproofing boats. Although the resin appeared to be a brittle solid at room temperature—something that would shatter into shards if hit by a hammer—it retained the properties of a liquid. Parnell knew that he could prove this to his sceptical students using a very simple experiment.

In May 1962, the fourth drop fell. There was no media interest, just some bemused staff . . . to note that a significant event had occurred.

JOHN MAINSTONE

1/8 SCALE FERMI REACTOR MODEL

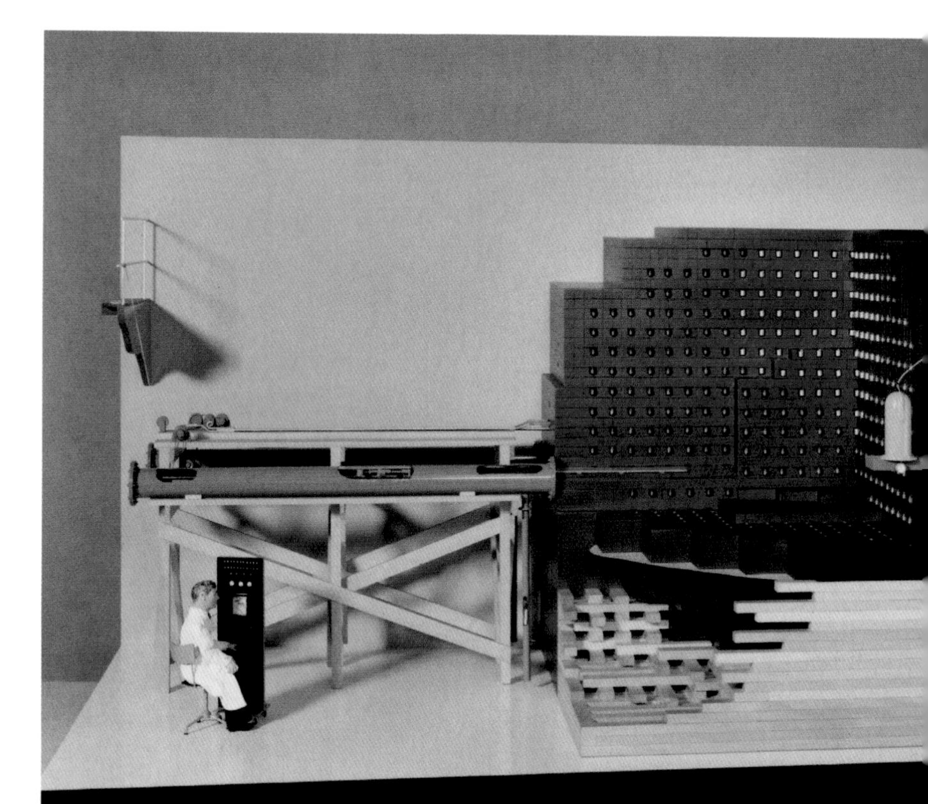

Fermi's experiment demonstrated that a self-sustaining nuclear chain reaction was not only achievable, but also that it could be produced with naturally occurring uranium. This was a crucial step towards the development of the nuclear bomb and its first test detonation in the US state of New Mexico in 1945. The experiment also paved the way for other experimental reactors of various types and designs, as well as for the development of nuclear power stations and nuclear-powered vehicles, ranging from aircraft carriers to space probes.

This model of the world's first nuclear reactor shows a cutaway of its interior structure and the central control rod known as the zip.

NUCLEAR FISSION
ENRICO FERMI
1942

The world's first nuclear reactor, dubbed Chicago Pile 1 because it was composed of a pile of graphite and uranium, was built by Italian physicist Enrico Fermi under the stands of the sports stadium at the University of Chicago in Illinois. It comprised 40,000 pure graphite bricks, with alternating layers of bricks containing uranium metal and/or uranium oxide. The graphite was used to slow the passage of neutrons from the uranium, thereby increasing the likelihood of them crashing into other uranium atoms and starting a chain reaction. A central control rod, covered in a sheet of neutron-absorbing cadmium, was used to hold the reaction at bay. Known as 'the zip', it was secured to a weighted pulley system so that when it was released, the rod would be pulled into the reactor and the reaction would be deadened. On the morning of 2 December 1942, a small group of scientists watched as the zip was eased out a little at a time and Fermi measured the progressive rise and levelling off of neutron activity. When criticality was reached that afternoon, the neutron counter moved faster and faster, refusing to level off and therefore indicating a nuclear fission chain reaction. After twenty-eight minutes, the zip was reinserted to absorb the neutrons and end the reaction.

By 1942, many scientists were working on the Manhattan Project, pursuing different areas of research to find a way to build a nuclear bomb. Fermi knew the key to progress was to create a self-sustaining nuclear chain reaction.

Fermi (1901–54) was renowned as both a theoretical and an experimental physicist who was an expert on the neutron. He won the Nobel Prize in Physics in 1938 for his nuclear reaction bombarding uranium with slow neutrons. Instead of producing heavier elements through fusion as he had hoped, Fermi's fission reaction split uranium into two radioactive isotopes of a smaller element.

MICROWAVE COOKERY

PERCY SPENCER

1945

In the mid 20th century, the ways in which people could cook in the domestic kitchen saw some major developments. Some of the new devices on the market at this time, such as the domestic pressure cooker, were based on long-established science, but one featured a surprising new use for the magnetron, a device that had been developed more recently.

In the mid 1940s, US engineer Percy Spencer was employed at defence contractor Raytheon, where he focused much of his attention on the magnetron. This device produces a beam of microwaves, which were a key component of early radar machines. One day, Spencer was working on a live magnetron circuit: power was flowing into the magnetron and microwave radiation was spewing out. During the experiment, he noticed that the peanut butter chocolate bar in his shirt pocket had melted. Although this melting effect had been observed previously, Spencer decided to explore the phenomenon and designed a further series of experiments. He brought a variety of foodstuffs into the laboratory in order to test what effect the magnetron would have. Corn kernels popped into popcorn in the face of the microwaves, and Spencer and his colleagues also discovered that eggs will explode if you fire microwaves at them. Before long, the team worked out that the microwaves were exciting the water molecules in the food, making them rotate. After receiving energy from the microwaves, the molecules pass on their energy by nudging neighbouring molecules into rotation, like the gears of a clock. The more rotating molecules present, the hotter the food.

Spencer (1894–1970) acquired most of his knowledge of electronics while serving in the US Navy. After joining Raytheon, he earned a Navy Distinguished Public Service Award—the highest award that the service can bestow upon a civilian—for finding a way to simplify the mass production of radar equipment. After the success of his invention, he rose to become a board member at Raytheon.

Early microwave ovens . . . would take 20 minutes to warm up . . . but they were 10 times more powerful than anything you can buy today.

ROD SPENCER, PERCY'S GRANDSON

Water absorbs microwave radiation, turning it into heat, but the same radiation bounces off metal. The insides of all microwaves ovens are lined with metal in order to trap the microwaves in the oven.

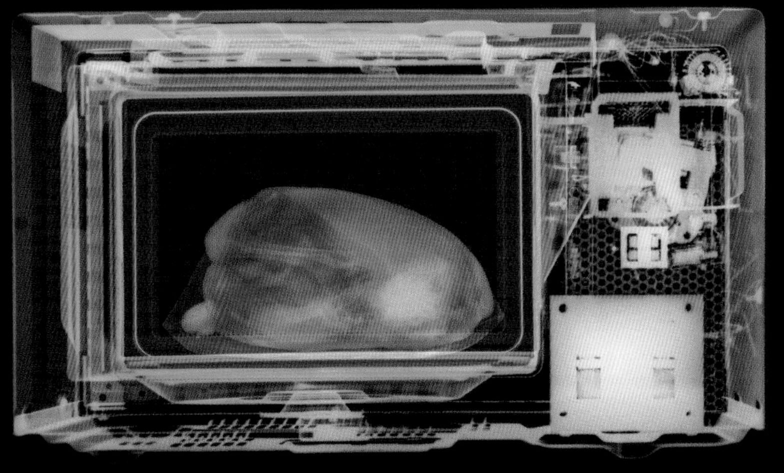

?

Raytheon was quick to realize the potential of, and to patent, various aspects of Spencer's discovery, but its initial attempts to create a commercial product for the domestic market flopped: the 'Radarange' was too big and expensive to catch on. Although the science of the microwave oven was sound, it found commercial success only when it was redeveloped into a smaller, less powerful device, like the one in use today.

?

The establishment of carbon-14 dating revolutionized not only the field of modern archaeology, but also whole areas of historic and scientific research. Also known as radiocarbon dating, it can be used to authenticate artworks and other relics. Although the method has its limitations and it is applicable only to organic matter within a limited time frame, it can be used for much of recorded human history.

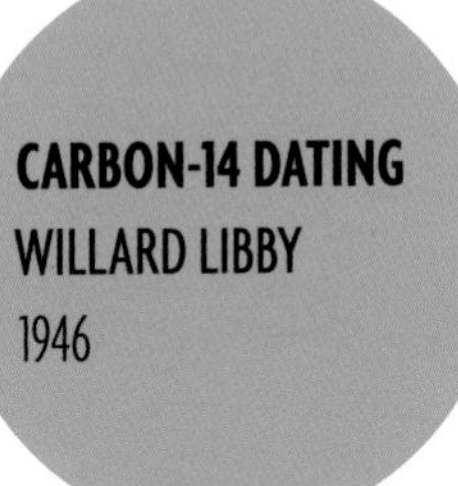

CARBON-14 DATING
WILLARD LIBBY
1946

Once US chemist Willard Libby ascertained that, in theory, all organic matter contained a 'nuclear clock' of radioactive carbon-14, he set up various experiments to investigate this idea. His first was to test fossil ıethane against newly produced methane from the ewage works of Baltimore, Maryland. As expected, the ossil sample had no carbon-14, because the isotope ompletely decomposes in around 50,000 years, and only ıe fresh methane had carbon-14. Libby's next task was to alibrate his dating technique by correlating the standard ecay curve of carbon-14, as it was known in 1946, to the vels found in organic test objects of known ages. This ould require what Libby termed a 'curve of knowns': a ollection of organic objects from throughout human istory with known providence, and therefore a known ge, stretching back to 50,000 years ago, or as close as as possible. He was surprised to find that the oldest em available dated back to the First Dynasty of ancient gypt, only about 5,000 years ago. However, historians ound the world were able to provide him with enough bjects of different dates, including bread from Pompeii '9 CE), to calibrate successfully the accuracy of his dating chnique. Libby went on to date other historic sites efore writing up his findings for publication.

Carbon-14 dating began with a simple realization that the level of radioactive carbon-14, present in all living things, would degrade over time from the point at which that living thing died. Therefore, by measuring the degradation of the carbon-14 in the organic material against that in the atmosphere, it would be possible to assess roughly the age of that material.

Libby (1908–80) recognized the importance of atomic research early in his career and participated in the Manhattan Project (1942–45) on uranium enrichment. After World War II, he turned his attention to various methods of dating substances through their radioactivity. In addition to his work with carbon-14, for which he won the Nobel Prize in Chemistry in 1960, Libby developed a similar technique for dating water by measuring its tritium content, a method sometimes used for establishing the true age of wine.

Radioactive carbon-14, with its half-life of 5,730 years, was first discovered in 1940 by US physicists Martin Kamen and Sam Ruben.

Libby kept his involvement in the Manhattan Project secret from his wife until the nuclear attack on Hiroshima, at which point he brought home a stack of newspapers and told her 'this is what I've been doing'.

MAPPING THE OCEAN FLOOR

MARIE THARP AND BRUCE HEEZEN

1948–77

Marie Tharp began her ocean floor-mapping odyssey in 1948 when she joined the Lamont Geological Laboratory (now the Lamont-Doherty Earth Observatory) at Columbi University. At the time, women were not allowed on board survey vessels, so it fell to her colleague, Bruce Heezen, to undertake the expeditions on the university's research ship *Vema*. Durir the surveys, he took advantage of the recent advances ir sonar (sound navigation and ranging) equipment, which sends sound signals through the water to bounce off the ocean floor and back up to the surface. Each sounding captured data on the seabed. Tharp analysed thousands of individual soundings that Heezen recorded, and also interpreted data from other organizations, including the US Navy and United Kingdom Hydrographic Office, as well as all the available earthquake data. She then plotted them, longitude degree by latitude degree, into a physiographic map of the ocean floor. The resultant detailed map of the North Atlantic Ocean floor showed the Mid-Atlantic Ridge running from north to south. More importantly, the map indicated that a central rift ran its entire length and divided the ridge.

Before their pioneering ocean-mapping collaboration, US geologists Tharp (1920–2006) and Heezen (1924–77) worked together to identify downed aircraft using images from deep sea cameras. The pair received the National Geographic Society Hubbard Medal for their mapping work in 1978, a year after Heezen's death. In 1999 a USNS oceanographic survey vessel was named the *Bruce C. Heezen* in his honour. Tharp continued to work into the mid 1990s and received the first Lamont-Doherty Earth Observatory Honors award for her achievement in earth sciences. She donated her notes and map collection to the US Library of Congress.

Techniques for mapping the ocean floor have progressed immensely since 1977. This map of the North Polar seabed was captured using satellite imaging in addition to the sonar techniques employed by Tharp and Heezen. It is the result of a project that began in 1997.

In the late 19th century, the main method used to assess the depth of the seabed was to lower a weighted rope out of the side of a boat until it began to slacken. This was laborious and it was performed mainly in coastal waters to aid navigation. However, the advent of sound-based techniques to measure depth and the eventual development of sonar in World War II revolutionized this task and scientists began to pay more attention to the ocean floor.

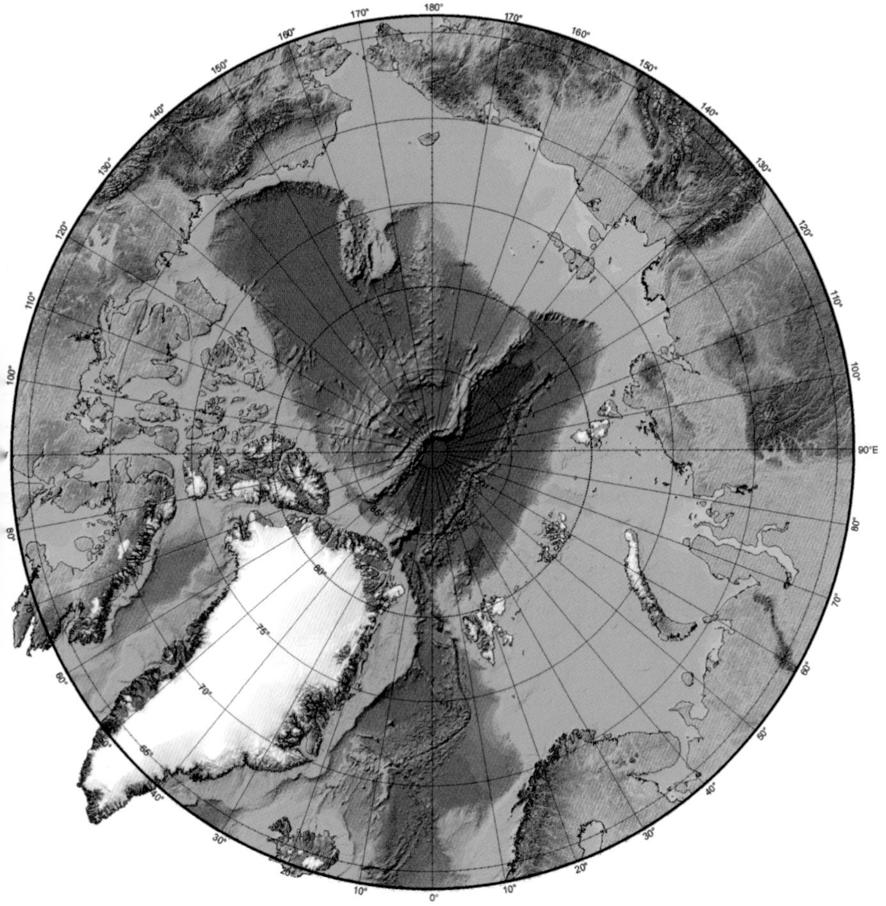

Tharp and Heezen mapped the last unmapped area on the face of Earth. Their first map of the North Atlantic Ocean floor was published in 1959 and—in conjunction with the work of US geologist Harry Hess, who measured gravity anomalies in the Pacific Ocean—it helped to establish what would later become the theory of plate tectonics. Tharp and Heezen went on to map the whole of the ocean floor, and this map was published in its entirety in 1977.

I had a blank canvas to fill with extraordinary possibilities . . . [It was] a once-in-the-history-of-the-world opportunity for anyone, but especially for a woman in the 1940s.

The early stages of the Manhattan Project (1942–45) explored a variety of methods to produce an energetic nuclear reaction. Although fusion reactions, in which hydrogen fuses into helium, were considered impractical for development during World War II, the early years of the Cold War led to renewed interest.

THERMONUCLEAR FUSION

EDWARD TELLER

1952

When US physicist Edward Teller began lobbying for the development of a fusion bomb—a device that releases energy through the fusion of smaller elements, such as hydrogen, rather than via the fission (breaking apart) of heavier elements, such as uranium—he knew that the technical challenge would be how to ignite the fuel. Indeed, his theories needed to be refined by mathematician Stanislaw Ulam before they could be implemented in the thermonuclear bomb design that is now known as the Teller-Ulam configuration. Physicist Richard Garwin was then assigned to design a device to prove that such a bomb would work. The exact details of the device, referred to as the 'Sausage', remain classified, but in essence a primary fission bomb was used to detonate a secondary fusion bomb by providing intense heat and compression to a flask of liquid deuterium, with a plutonium rod at its centre to act as a spark plug. Built on Elugelab in the Pacific Ocean, the Sausage had refinements to increase the heat and pressure on the deuterium fuel in order to ensure it would produce an effective fusion reaction. Its cooling systems, required to keep the deuterium fuel in liquid form before detonation, made the device bulky; it was the size of a shed and weighed some 50 tonnes (55 tons) . On 1 November 1952, the device was detonated, causing the first megaton nuclear blast.

Widely considered to be the 'father of the hydrogen bomb', Teller (1908–2003) was an early advocate of the fusion bomb. However, disagreements with Ulam led him to bow out of the final design and detonation of the device in the Ivy Mike test.

Code named Ivy Mike, the successful detonation of the Sausage, its high megaton radioactive yield and the mile-wide crater it created proved how effective thermonuclear weapons based on fusion would be. Development work to design a practical version of the bomb began soon after. In addition, as part of the data collection process, specially equipped F-84 fighter/bomber planes were sent to take samples from within the mushroom cloud two hours after detonation. Among the readings, scientists detected two synthetic elements, ones that are too unstable to be found naturally on Earth.

The mushroom cloud from Ivy Mike reached well in excess of 30,480 metres (100,000 ft) in height and was clearly visible to all those on the USS *Estes* observing the first megaton test shot.

Physicist John von Neumann (1903–57) was a member of the Atomic Energy Commission and one of the key contributors to report on Operation Cue. He is also credited with originating the theory of mutual assured destruction (MAD) in nuclear policy.

H minus one minute. Put on your goggles. Observers without goggles must face away from the blast.

Operation Cue served a valuable dual purpose. Firstly, it tested the resilience of US infrastructure and home construction to nuclear attack, and secondly it reassured the US public that, by taking precautions and building reinforced shelters, they could survive a nuclear attack. However, the film of the test pointed out that the test bomb was far weaker than the multimegaton nuclear weapons that were available in the mid 1950s. A real nuclear attack would be likely to cause far more devastation.

The rows of mannequins, lined up in the open air to face the nuclear blast, suffered quite badly, with limbs blown off and patterns from the dresses burned onto the mannequin beneath.

OPERATION CUE

ATOMIC ENERGY COMMISSION

1955

'Survival Town' was the name of the small collection of buildings constructed at the Nevada Test Site in Yucca Flats on behalf of the US Federal Civil Defense Administration to better test the resilience of standard US infrastructure to a nuclear attack. The town, located about a mile from where the nuclear device was detonated, included a radio station, electrical substation, five types of residential housing and a variety of other buildings. They were all fully furnished with modern appliances and decor and well-stocked with a variety of tinned and otherwise preserved food. Further perishable food was buried just below the soil to be tested for nuclear contamination after the detonation. As a finishing touch, the town was populated with families of mannequins. They wore the latest fashions and were placed in everyday poses, or as close an approximation as was possible, both in the buildings and outside. Film cameras were positioned throughout the town to record exactly how the blast hit each structure. In the early hours of 5 May 1955, a 30-kiloton nuclear device dubbed Apple-2 (roughly twice as powerful as the bomb dropped on Hiroshima in 1945) was detonated. Approximately 6,000 people witnessed the experiment, either through heavily tinted goggles or with their backs turned to the explosion. After twenty-four hours, participants were allowed back into the town to assess the damage. Although a number of the buildings had been destroyed, special reinforcements had left other buildings more intact than in previous tests.

Many manufacturers supplied the goods to furnish Survival Town free of charge; US retailer JCPenney provided the clothing.

In the 1950s, the US government conducted many nuclear tests, primarily focused on developing weapons. However, in 1953, two houses were built near a detonation site to assess how resilient they would be to a nuclear blast. Both were destroyed completely; more tests were needed.

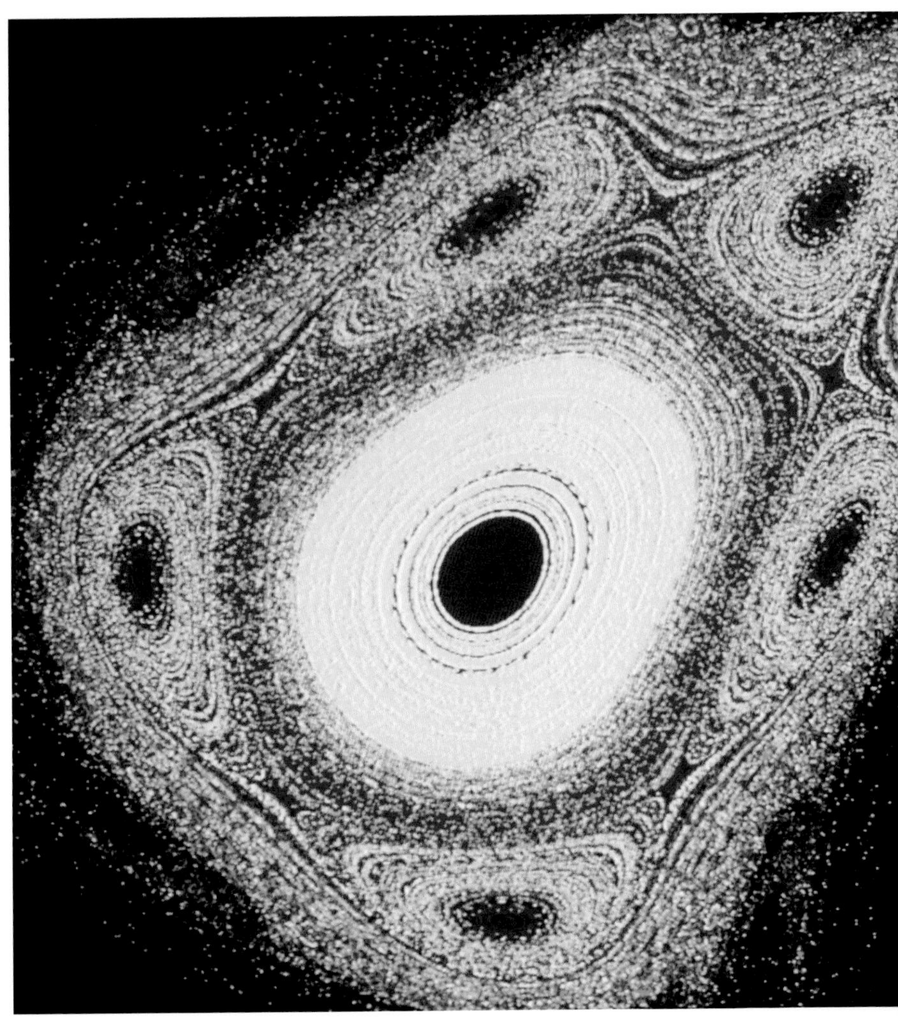

?

Lorenz's findings—that small changes in initial conditions can rapidly lead to widely divergent results—spawned chaos theory. In terms of weather prediction, the chief implication of chaos theory is that it is unlikely that we will ever be able to produce accurate long-term forecasts due to our limited knowledge and limited ability to determine the exact initial conditions for any given day.

i

The Hénon map is a means of depicting patterns that develop out of chaotic systems, such as stellar globular clusters.

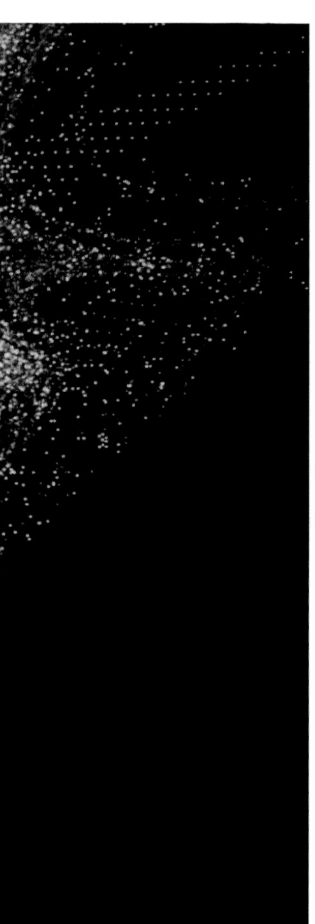

Humans have always sought long-term weather predictions. However, despite an increasingly accurate record of past weather, we are still unable to predict the weather in detail for more than a short period of time. In the 1960s, Lorenz identified part of the reason behind this uncertainty.

CHAOS THEORY
EDWARD LORENZ
1961

To some extent, each day's weather can be seen as an experiment that is rerun daily. This was the approach taken by US meteorologist Edward Lorenz in 1961 when he ran a simple weather simulation, on an early computer, to produce predictions based on a limited set of initial conditions. He was attempting to repeat one model in particular, using the same initial values for all circumstances. Although he expected to obtain the same results for each model, he noticed that the results of the two predictions diverged widely. He realized that the only difference between the two versions of the experimental models was a single field in which he had entered a value to six decimal places initially but subsequently shortened that value to three decimal places. This tiny change in initial conditions of less than 0.01 per cent resulted in massive changes to the prediction.

In 1972 Lorenz's talk for the American Association for the Advancement of Science was titled 'Does the Flap of a Butterfly's Wings in Brazil Set off a Tornado in Texas'.

Lorenz (1917–2008) was a weather forecaster in the US Army Air Corps during World War II. There, he developed an interest in meteorology, the interdisciplinary branch of science that led him to develop the tools of chaos theory. For this feat, he was awarded the Kyoto Prize in Basic Sciences in 1991.

The impact of humans on the world's climate could be seen as the largest ongoing experiment in the world. In order to demonstrate the climate history of our planet, we need to look back further than the 150 years of globally recorded weather data and seek evidence of the planet's past temperature and atmosphere, frozen in ice.

CLIMATE CHANGE

WILLI DANSGAARD AND CLAUDE LORIUS

1964–present

In 1952 Danish paleoclimatologist Willi Dansgaard first noticed a correlation between air temperature and the amount of oxygen-18 in precipitation. He theorized that the colde the temperature the more isotopes of oxygen-18 would condense as snow or rainfall. Over the following years, he tested his theory with ice samples supplied by the Atomic Energy Agency, but it was not until 1964 that he managed to gain access to a deep ice core in Greenland. In collaboration with th US military, which drilled some 1,390 metres (4,560 ft) int the ice, Dansgaard was able to analyse samples and to plot global temperatures going back 11,000 years. The results showed a period of intense glaciation took place more than 10,000 years ago. A year later, French glaciologist Claude Lorius developed another method o obtaining climate data from ice. The revelation came to him when he placed a chunk of Antarctic ice in a glass of whisky during a winter expedition. Watching the ice melt in his glass, Lorius noticed trapped bubbles of air within the frozen water. He realized that the air in these bubbles was a 'reliable and unique indicator . . . of the composition of the air' for the period when the ice was first formed. It therefore contained valuable informatior that would greatly increase our knowledge of climate conditions hundreds and thousands of years ago.

Current greenhouse gas levels are unprecedented, higher than anything measured for hundreds of thousands of years, and directly linked to man's impact on the atmosphere.
CLAUDE LORIUS

Dansgaard (1922–2011) and Lorius (1932–) shared the Tyler Prize in 1996 (with Hans Oeschger) in recognition of their contributions to climate science. Lorius also served as the director of the National Centre for Scientific Research in France and won the Balzan Prize for climatology in 2001.

This ice thermometer from 1857 was designed to measure the temperature of Alpine glaciers. In order to obtain data from earlier ages, it is necessary to drill down into ice that has accumulated over many thousands of years.

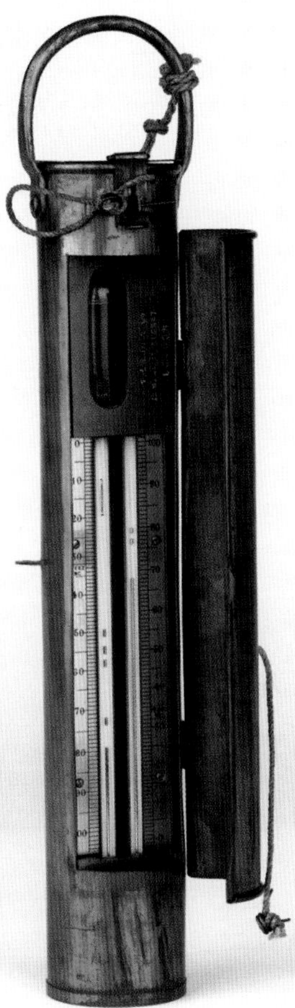

By analysing the air bubbles and temperature records on his expedition to the remote Soviet Antarctic Vostok station, Lorius and the glaciologists who followed him have been able to establish a climate record for a period of roughly 150,000 years, covering all climate cycles of the late Quaternary period. This has shown a strong correlation between rising temperatures and rising levels of carbon dioxide and methane in the atmosphere. It has also indicated that the current levels of these gases are higher than at any point recorded previously.

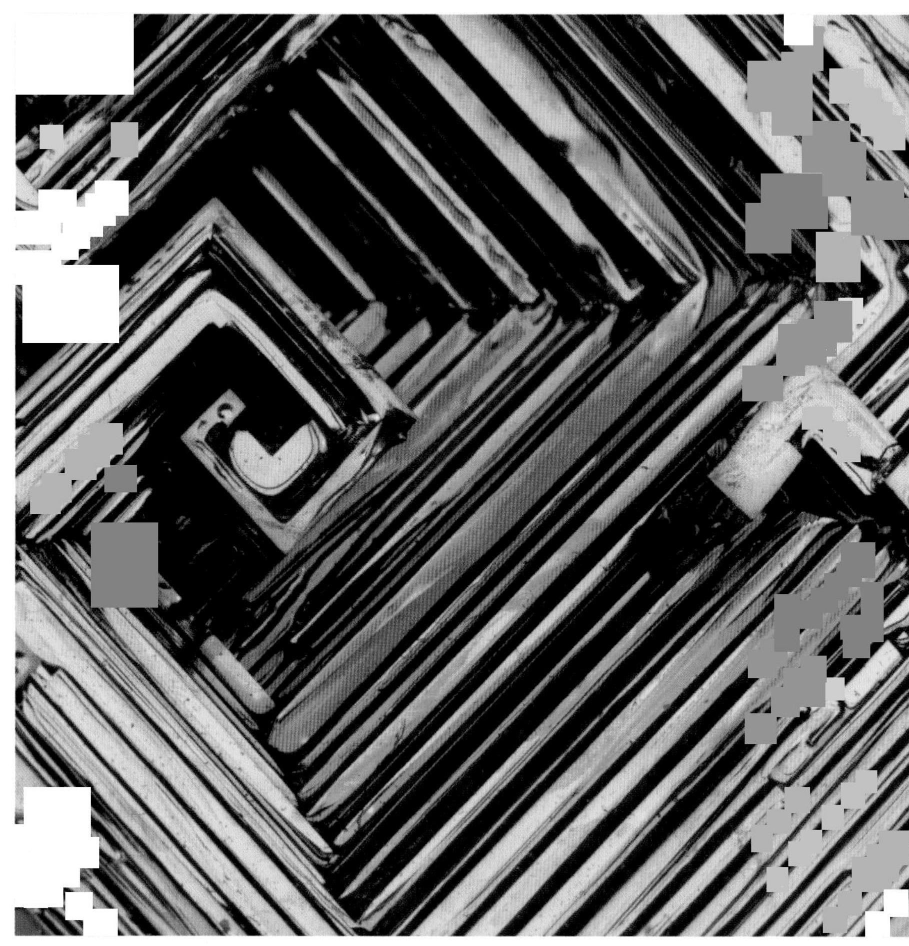

Although the experiment was a success, it demonstrated the huge cost of creating gold by smashing together atoms. The team of scientists estimated that the expense incurred for the use of the particle accelerator for the day of the experiment would have been in the region of US$40,000. It produced only a few atoms of gold, the most stable of which could not be detected. Therefore, the experiment serves to underline that although many things are possible scientifically, they are not necessarily economically viable.

The rectangular crystalline structure of bismuth can be created only in laboratory conditions.

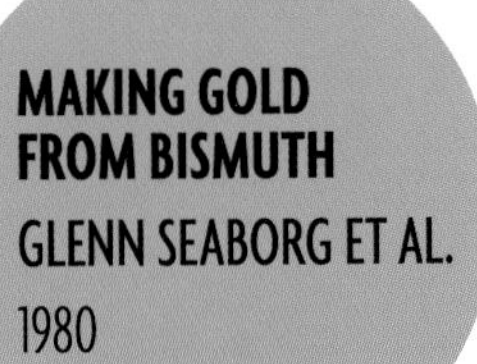

MAKING GOLD FROM BISMUTH

GLENN SEABORG ET AL.

1980

At the Lawrence Berkeley National Laboratory, California, a group of scientists devised an experiment using the institution's Bevalac particle accelerator. They fired beams of carbon and neon nuclei, at a peed approaching that of light, at small foils of the etal bismuth. They used bismuth because it has only ne stable isotope rather than the four that are found 1 lead. The aim of the experiment was to smash off our protons and enough neutrons from various bismuth toms to turn them into atoms of various isotopes of old. Searching through the debris of the collisions, the cientists found atoms of isotopes of gold, each with he requisite 79 protons and anywhere from 111 to 120 eutrons. However, the amount of gold created was so mall that detecting it was only possible by measuring he radiation from the radioactive isotopes of gold as hey decayed over the course of a year. Ironically, any toms of gold's stable isotope 197—composed of 79 rotons and 118 neutrons, the type used for wedding ings, for example—that might have been created in the xperiment would not have been possible to detect hrough its radiation. Furthermore, there would have een too few atoms to detect through any other means.

Of the five scientists credited with creating gold from bismuth in this experiment, the most senior was US nuclear chemist Glenn Seaborg (1912–99). He won the Nobel Prize in Chemistry in 1951, which he shared with fellow chemist Edwin McMillan 'for their discoveries in the chemistry of the transuranium elements'. Seaborg was credited in the discovery of ten elements, one of which was named after him. He also served as head of the Atomic Energy Commission from 1961 to 1971. In this role, he played a prominent part in orchestrating the Partial Nuclear Test Ban Treaty in 1963, a first step in the process to end the proliferation of nuclear weapons.

Transforming base metals, such as lead, into gold has presented an ongoing challenge to alchemists since ancient times. However, during the 20th century, as scientists developed a much better understanding of chemistry and the structure of various chemical elements, a number of them began to consider the possibility of utilizing modern techniques to accomplish the task.

It would cost more than one quadrillion dollars per ounce to produce gold by this experiment.

Marine biologists use Aquarius Reef Base to make surveys of the local reef coral and sponge populations. Some 600 peer-reviewed science papers have been credited to these scientists, but perhaps the most notable mission to take place at Aquarius was when NASA used it in 2001 as a training facility for its astronauts who were heading for the similarly hazardous environment of the International Space Station.

Located 19 metres (62 ft) below the surface, Aquarius Reef Base has become part of the subaquatic scenery.

In the 1960s, the US Navy learned that divers can perform more dives if they are kept at the same pressures that they experience at depth. This technique of diving from a pressurized environment is called saturation diving and its discovery enabled scientists to develop facilities for long-term underwater living and research.

UNDERWATER LIVING

FLORIDA INTERNATIONAL UNIVERSITY

1986–present

First deployed in the waters off the US Virgin Islands in the 1980s, and refitted and repositioned in the Florida Keys in 1993, Aquarius Reef Base has been operated by Florida International University since 2012 as a base for long-term aquatic experiments and observations. The underwater facility, which is kept at the ambient pressure of the surrounding sea to facilitate saturation diving, can accommodate up to six scientists and technicians. It comprises a living area—with bunks, work area and kitchen facilities, as well as high speed internet access—a middle airlock and a 'wet porch', a pressurized room that provides access to the sea through an aperture beneath the base. Missions to Aquarius typically last ten days, but some have continued for as long as a month. 'Aquanauts' can dive for up to nine hours a day, as opposed to the two-hour limit associated with most surface-based dives.

The world record for the longest time spent living underwater is sixty-nine days, set by NASA researcher Richard Presley in 1992 at Jules' Undersea Lodge.

Over a series of experiments between 1957 and 1969, US physician and naval captain George Bond (1915–83) helped his service to develop the technique of saturation diving, for which he received the Navy Commendation Medal. He also served as the senior medical officer of the SEALAB project.

The San Andreas Fault was first identified by British geologist Andrew Lawson in 1895, many years before the theories of plate tectonics were advanced. The fault's significance only became clear a decade later when it produced an earthquake that caused catastrophic damage to the city of San Francisco.

PREDICTING EARTHQUAKES

EARTHSCOPE

1990s–present

Set up in the late 1990s, EarthScope is an umbrella organization that combines the work of numerous scientists in ongoing experiments, conducted at three main geological observatories: the US Array, the Plate Boundary Observatory and the San Andreas Fault Observatory at Depth (SAFOD). The geologists are keenly aware of the importance of fault lines and their link to seismic activity, and they are collecting data that can be used to map the dynamics of the geology of the North American continent. SAFOD is focused on the processes of the San Andreas Fault. Planned since 1992, the project commenced in 2003 and involved drilling two cores to a depth of 3 kilometres (1.8 mi). These were used to collect rock and fluid samples from across a particularly active stretch of the fault. The geology and composition of these cores was then studied separately. In the second phase of the experiment, which started in 2008, geophysicists deployed the 'observatory at depth', in the form of various sensors and monitors, including a passive electromagnetic coil that could detect any electromagnetic waves radiated by the earthquake source. Together, these devices provide important data on the processes of micro earthquakes from within the San Andreas Fault. The purpose is to be able to predict and interpret future earthquakes.

Lawson (1861–1952) was employed as a geology professor at University of California, Berkeley, at the time of the earthquake in 1906. He went on to co-author and edit the official report on its causes. He subsequently mapped the full extent of the fault line.

Since they were first brought together, the EarthScope observatories have generated significant data, all of which is archived and freely available. SAFOD alone has been cited in a total of 2,390 papers as of December 2015, but many of the sensors in the observatory have now failed and need to be replaced. Despite this ongoing research, it remains unclear how much advanced notice of the next big earthquake, be it in the San Andreas Fault or elsewhere, the various EarthScope observatories will be able to provide.

The San Francisco earthquake of 1906 was one of the most devastating in US history. Figures for the death toll vary, but more than 25,000 buildings were destroyed, rendering their inhabitants homeless. It is hoped that the more seismological data we have access to, the greater our preparedness for when the next quake strikes.

ISOLATING GRAPHENE
ANDRE GEIM AND KONSTANTIN NOVOSELOV
2004

During an informal 'Friday night experiment' session, physicists Andre Geim and Konstantin Novoselov were exploring the electrical properties of flakes of carbon graphite. They decided to see if they could make the flakes even thinner by using ordinary adhesive tape. They performed what Novoselov later termed 'micromechanical cleavage'—pulling a few layers of graphite from the block onto the tape—and they were able to deposit as little as a single layer of graphite from the tape onto other surfaces. They dubbed this two-dimensional crystalline graphite material graphene, and soon found that it was an excellent conductor of electricity, despite its extreme thinness. This was demonstrated when they used a few layers of graphene to make a metallic field-effect transistor. After performing a number of experiments on the electric field effects of graphene, Geim and Novoselov concluded that it could be the ideal material for the construction of nanoscale transistors and that other applications might be even more exciting. Subsequent experiments have found that graphene is incredibly strong and flexible—about one hundred times stronger than steel by weight—and it can be stretched. It is also transparent and the best conductor of heat discovered to date.

Geim (1958–) and Novoselov (1974–) shared the Nobel Prize in Physics in 2010 for their 'groundbreaking experiments regarding the two-dimensional material graphene'. The pair have both been recognized as fellows of the Royal Society, but only Geim has been awarded its prestigious Copley Medal for his graphene work. Geim also holds the distinction of being the only individual to win both the Nobel and Ig Nobel prizes, the latter of which he won for his work levitating a frog in a powerful electromagnet.

The carbon that forms graphene is renowned for its ability to bond with a wide variety of materials. In forms such as the carbon fibre depicted here, it is used to strengthen everything from aircraft hulls to sports equipment.

In 2013 the University of Manchester estimated that the global graphene manufacturing industry was already worth more than $10 million, with a further $2.4 billion made available for research.

The element carbon has long been known for its remarkable properties. It serves as the main element in all known life and it can bond with a variety of other elements to form more than 10 million known compounds. Carbon can also bond with itself to create different carbon structures, including diamond, graphite, carbon nanotubes and graphene.

Although graphene was discovered during a light-hearted experiment using everyday objects, its exciting properties—coupled with the rapidly decreasing cost of its manufacture—have helped to make it an extremely popular area of research. Between 2009 and 2013, the annual worldwide production of graphene saw a 1,608 per cent increase. And between 2008 and 2012, various companies and individuals registered more than 7,000 patents for applications based on graphene. The first products to reach the market to date include a graphene-based ink for printing thin flexible electrical circuits, touch screens for portable electronic devices and even tennis rackets. However, many more are expected in the not too distant future.

Frankly speaking, I value both my Ig Nobel Prize and Nobel Prize at the same level.

ANDRE GEIM

CHAPTER FIVE
THE UNIVERSE

As writer Douglas Adams observed in *The Hitchhiker's Guide to the Galaxy* (1978), 'Space . . . is big. Really big. You just won't believe how vastly, hugely, mind-bogglingly big it is.' And so science fact bears out this statement of science fiction. Some experiments demonstrate the physical laws of the universe, such as the wave/particle duality of light, whereas the various experimental proofs of Einstein's theories of relativity have led to Edwin Hubble's observations that the universe is indeed expanding. Since scientists have been exploring space via manned and unmanned missions, these experiences have served to emphasize the immense scale of the universe and its many remaining mysteries, including the as-yet-unknown nature of dark matter and dark energy.

‹ Detecting the Higgs Boson (see p. 216)

Like Isaac Newton, Young (1773–1829) was a polymath with an active interest in multiple areas of the arts and sciences. He wrote extensively on a variety of subjects for *Encyclopaedia Britannica* and was a member of both the Royal Society and the Royal Institution. He was also an Egyptologist and a linguist who helped decipher the text on the Rosetta stone. Young even developed his own system for tuning musical instruments.

The experiments I am about to relate . . . may be repeated with great ease, whenever the sun shines.

These illustrations of various optical phenomena are from Young's book *A Course of Lectures on Natural Philosophy and the Mechanical Arts* (1807), a collection of the lectures he delivered on physics.

INTERFERENCE OF LIGHT

THOMAS YOUNG

1803

When British physician and physicist Thomas Young first presented his experiment before members of the Royal Society in 1803, he called it simply an 'experimental demonstration of the general law of interference of light'. His intention was to use a thin piece of card to divide a beam of sunlight. Young employed a mirror to direct the sun's rays into the lecture room through a hole in its window shutters. He then covered the hole with a thick piece of paper containing a single pinhole. This created a cone of sunlight that projected onto the wall. When Young inserted his card edgeways into the beam of light, he split it into two beams divided by an area of shadow. Coloured light was visible at the fringes of the shadow, and within the shadow were parallel fringes, or bands, of white light that appeared brightest in the centre of the shadow. Blocking the light on either side of the card caused the fringes to disappear. This demonstrated that the bands were caused by diffraction, due to the wave nature of light. Later versions of the experiment created two interfering beams, when a monochromatic light was shone through a piece of card that contained two slits. These slits gave the experiment the name 'double slit', by which it is known today.

Italian priest and physicist Francesco Grimaldi coined the word 'defraction' in the mid 17th century to describe the effect of two beams of light interacting. However, by the start of the 19th century, scientists widely believed, in line with the work of Isaac Newton, that light was composed of particles that travelled in straight lines, bouncing off objects at regular angles.

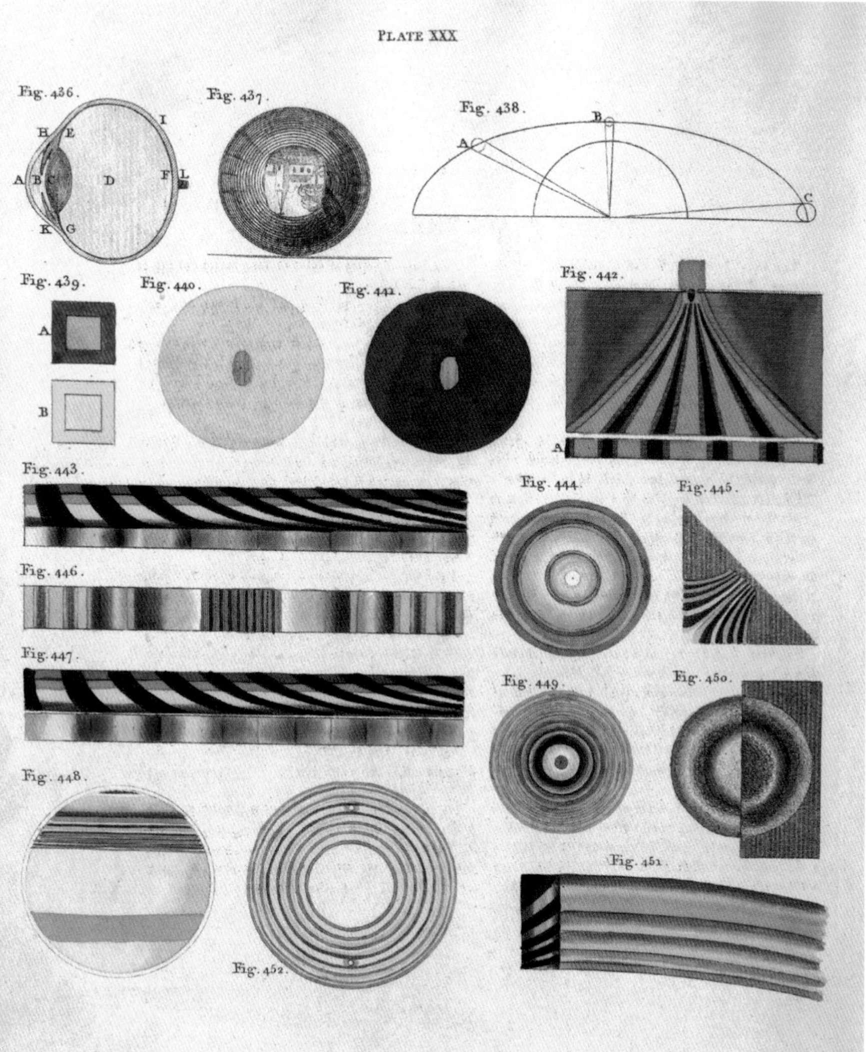

Young's experiment established the principle of interference of light. Although he designed it to demonstrate how light operates as a wave, subsequent studies have shown that light operates as both a wave and a particle. Physicists including Niels Bohr, Albert Einstein and Richard Feynman have used modified versions of Young's experiment to explore the quantum mechanical effects of this wave/particle duality.

Ballot (1817–90) is best known for his work in the field of meteorology, giving his name to Buys Ballot's law in 1957. This law is used to determine areas of high and low atmospheric pressure in relation to the direction of the wind.

Ballot had one main recommendation for anyone replicating his experiment: be sure to hire disciplined professional musicians.

After the experiment, Ballot was able to correct some errors that he had spotted in Doppler's original description of the theory and to devise a formula. However, he did not come up with any practical application for its use. This task was left to subsequent generations of scientists, who learned to measure an object's location by applying Doppler's principles. The Doppler effect now has many applications, including in radar, radio astronomy and sophisticated weather tracking and forecasting.

Trains in 1845 were often both noisy and dirty, which makes the success of Buys Ballot's experiment all the more impressive.

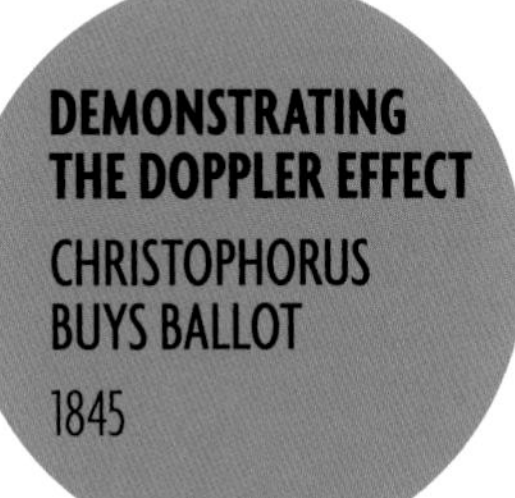

DEMONSTRATING THE DOPPLER EFFECT

CHRISTOPHORUS BUYS BALLOT

1845

The experiment conducted by Dutch meteorologist Christophorus Buys Ballot was unusual to say the least. He was interested in the relation of movement to wavelength, suggested by Christian Doppler, and decided to stage a public demonstration in order to test it—using sound waves rather than light waves. He convinced those in charge of the recently opened Utrecht to Amsterdam railway line to allow him to conduct an experiment using one of their trains. He then brought together a group of trumpet players, each of whom was able to discern a precise musical note simply by listening to it. Ballot placed one of the trumpeters in an open carriage on the back of the train and instructed him to play a single note, a G, when he gave the signal. The rest of the musicians were placed beside the track in three groups, roughly 400 metres (1,310 ft) apart. As the train travelled away from the stationary parties, Ballot signalled the musician on board to play and sustain his note. Those standing by the side of the tracks were then tasked with listening for the pitch of the note. On the return journey, the positions were reversed and the musician played the note on the ground while the listeners passed by on the train. The experiment was quite elaborate, so Ballot conducted it a number of times over two days: 3 June and 5 June 1845. Getting the train to travel at a constant speed was not easy, but at the end of the demonstration Ballot had collected enough data to prove the existence of what became known as the Doppler effect on sound waves.

Ballot's musicians played valved trumpets on the first day and louder natural trumpets on the second.

In 1842 Austrian mathematician Doppler developed a theory that light changes its colour, and therefore its frequency, depending on the relative speed of the source in relation to the observer. Three years later, Ballot demonstrated this effect using sound waves and a train.

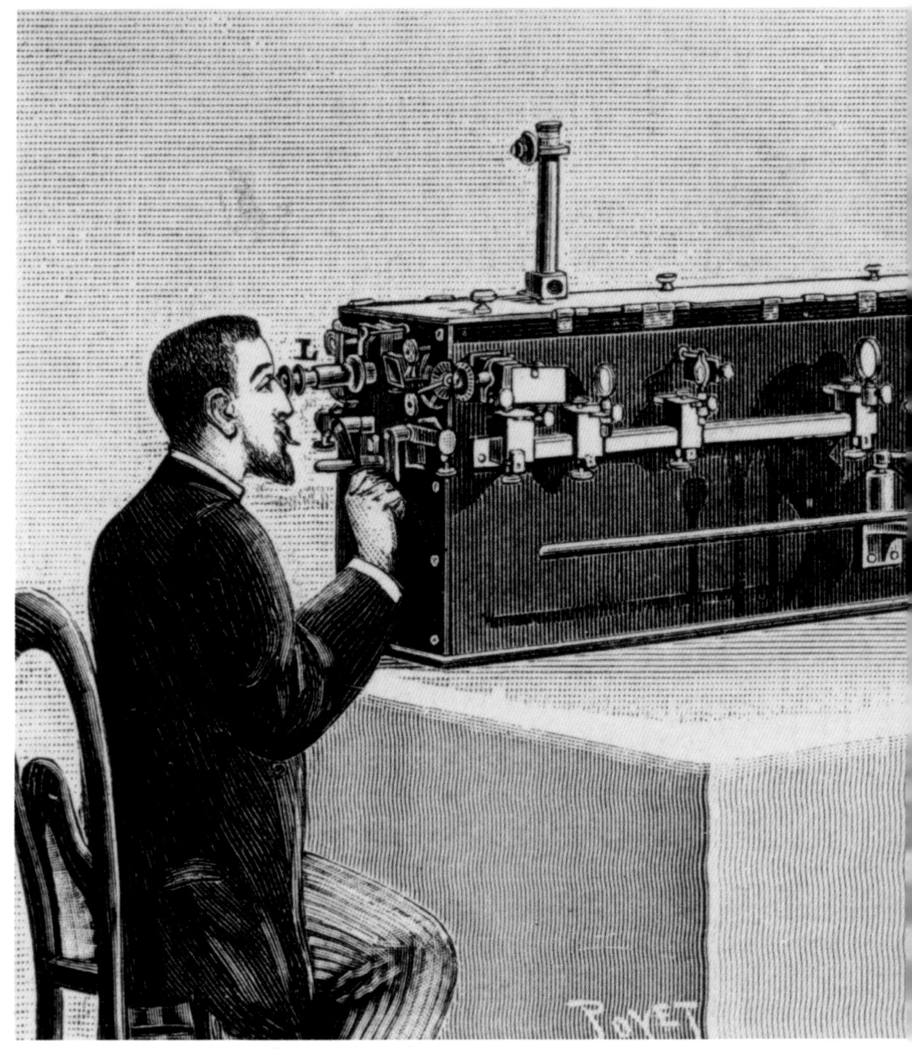

?

This experiment is somewhat unkindly considered the best-known 'failed' experiment in science. More accurately, it showed a null result: an absence of the expected effect. The scientists confirmed the data by repeating the test over subsequent months but the results remained the same. They provided some evidence that light might not need a medium through which to travel. However, it would not be until Einstein's special theory of relativity that the luminiferous ether was fully rejected.

Sound needs to travel through the medium of air and cannot travel through a vacuum. In the 19th century, many physicists believed that light must also travel through some kind of medium. This was dubbed luminiferous ether and testing its effect was the purpose of the Michelson-Morley experiment.

TESTING FOR LUMINIFEROUS ETHER

ALBERT MICHELSON AND EDWARD MORLEY

1887

This experiment was designed to measure the amount of drag the movement of light suffers as a result of travelling through the theoretical medium luminiferous ether as Earth orbits the sun. It involved an interferometer, which used a semi-silvered mirror to split light from a single source into perpendicular directions. This light was reflected by mirrors and recombined towards a light detector, where the interference patterns between the light from each side of the split could be observed. In order to increase the distance that the light travelled, and therefore highlight any measurable discrepancy, eight mirrors were employed to reflect the light backwards and forwards in each direction. Finally, the entire apparatus was mounted on a stone that was floating on mercury, with data recorded at sixteen positions as the experiment slowly rotated. Light, as it was theorized, would be slowed down by travelling against the flow of the ether as Earth passes through it. US physicists Albert Michelson and Edward Morley ran their experiment for three days but were surprised that the data they collected showed much less of a difference than the mathematical models for luminiferous ether had predicted.

More accurate apparatus was used in the Michelson-Gale-Pearson experiment in 1925 but achieved the same null result.

Michelson (1852–1931) and Morley (1838–1923) were both excellent experimental scientists. Michelson won the Nobel Prize in Physics in 1907 for his 'optical precision instruments' and Morley was awarded the Davy Medal by the Royal Society in the same year for 'his contributions to physics and chemistry'.

In addition to demonstrating Einstein's theory, Eddington (1882–1944) is well known for his astronomical observations and his calculations on processes underlying star formation. He won the Smith Prize at Cambridge University in 1907 for an essay on the proper motion of stars, and later became the director of the university's observatory. A long-standing member of the Royal Astronomical Society, he won its prestigious Gold Medal in 1924.

Einstein takes us straight to the root of the mystery, and he clears up one point which was misleading, if not actually wrong, in the older explanation.

This glass plate negative taken by Eddington during the solar eclipse in 1919 shows the stars near the sun, marked by horizontal lines, that had their light bent by the sun's gravity.

TESTING THE THEORY OF GENERAL RELATIVITY

ARTHUR EDDINGTON

1919

Although Albert Einstein had claimed that his theories provided an accurate model of physics, many scientists believed they required further testing before they could be considered proven. Part of Einstein's theory of general relativity proposed that the sun's gravity bends the light around it. In order to test this notion, British astronomer Frank Dyson proposed observing a solar eclipse. If Einstein were correct, light from stars close to the sun—not usually visible in daylight but visible during a total eclipse—would appear to shift slightly in their apparent location. The expedition to photograph the eclipse was led by British astronomer Arthur Eddington. He travelled to the African island of Principe for the eclipse on 29 May 1919, the next such event to occur in front of a dense star cluster. Another expedition was sent to Sobral, Brazil. Rain plagued the days leading up to the occasion, and clouds covered the sun for much of the 411-second eclipse. However, a brief ten-second gap in the clouds provided the opportunity for Eddington to take his photograph. Comparing his glass plate negative with another of the same star cluster, and also with the plates from Sobral, he proved that the sun did indeed bend the light of nearby stars, in line with Einstein's theory.

Before Einstein's theory of general relativity, the laws of classical physics, as delineated by British physicist and polymath Isaac Newton, were widely acknowledged as accurate. However, small discrepancies existed between Newtonian predictions and some measurable aspect of the solar system, such as the fact that the sun's gravity bends light.

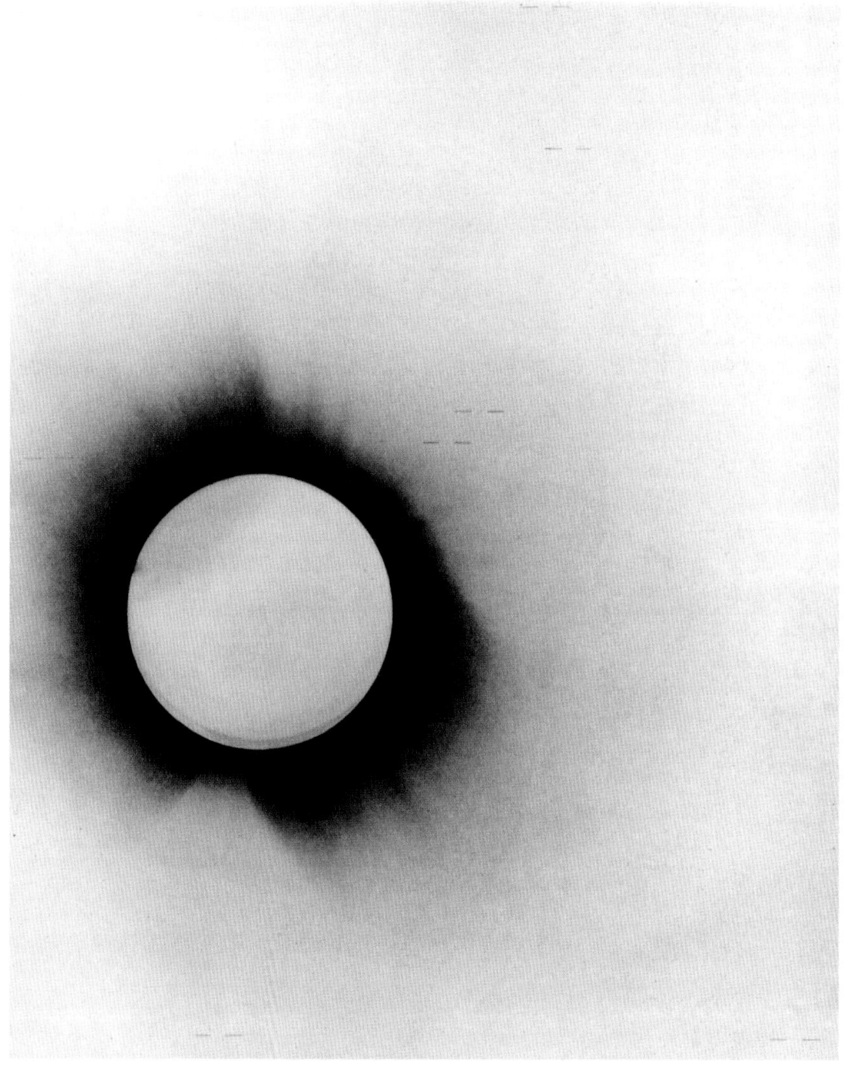

Eddington's observations were instrumental in gaining acceptance for the theory of general relativity and helped to overcome some of the prejudice that German scientists faced after World War I. Although the results were viewed sceptically by some, they have been reproduced a number of times with improved equipment, and both their accuracy and that of Einstein's theory have been borne out.

Hubble's Law is widely accepted as one of the cornerstones on which modern cosmology, the study of the origin and structure of the universe, is based. A number of different measurements of the universe rely upon it, including the Hubble Constant, which denotes the rate at which the universe is expanding. However, the precise measurement of this constant is still a matter of active investigation.

During the 1920s, Edwin Hubble worked at the Mount Wilson Observatory in California, using the powerful Hooker optical telescope. With a 2.5-metre (8 ft) aperture, the Hooker was the largest telescope in the world at the time. In 1923 Hubble began to observe Cepheid variables, a particular kind of regularly pulsating star, and was able to calculate the distance from Earth of the Andromeda Nebula, now known as the Andromeda Galaxy. He concluded that the Andromeda Galaxy was further away than the stars in the Milky Way, thus demonstrating that the universe was far larger than previously thought. Hubble went on to re-examine and classify celestial objects, previously known as nebulae, charting both their distance and their red shift. The latter is a shift of a galaxy's unique spectral emissions towards the red end of the spectrum as a result of wavelength being stretched out proportional to the speed of its recession from Earth. Hubble determined the distances to twenty-four nebulae and the red shift of forty-six. By comparing these two factors, he discovered a linear correlation between them, which indicated that the more distant objects were moving away the fastest. This relationship was part of the first experimental evidence that the universe is expanding.

Basing his work on the equations within Einstein's theory of general relativity, Belgian physicist George Lemaître proposed that the universe was expanding in a paper published in 1927, two years before that of Hubble. Although Lemaître's calculations were sound, his paper was not widely read or championed until after Hubble's observations had received recognition.

The discovery that the universe is expanding was one of the great intellectual revolutions of the 20th century.

STEPHEN HAWKING

Always passionate about astronomy, Hubble admires the Schmidt telescope at the Palomar Observatory in San Diego in 1949.

US astronomer Hubble (1889–1953) was awarded a wide array of honours in recognition of his discoveries, including the Gold Medal of the Royal Astronomical Society. Despite his many plaudits, the one he coveted most—the Nobel Prize—eluded him because the Nobel Committee did not recognize the science of astronomy at the time. Hubble lobbied on his own behalf and for other astronomers, but died before the committee began to recognize astronomers within the physics category.

DISCOVERY OF THE POSITRON
CARL ANDERSON
1932

On 2 August 1932, US physicist Carl Anderson was studying the effects of cosmic rays in his laboratory at Caltech, California. The main piece of equipment he used for this was a cloud chamber: a sealed glass cylinder containing an alcohol vapour that is kept in a constant state of super saturation, just above the point of condensation. Charged particles from cosmic rays, which are so tiny that they are usually invisible, flow through the chamber and can be detected by the distinctive cloud-like trail of condensation left in the wake of the particle as it ionizes the vapour. By surrounding the chamber with a sufficiently powerful consistent magnetic field, Anderson could observe the magnetic charge of the particles, which would spin or curve in opposite directions depending on their charge. As he continued his observations, Anderson was surprised to capture in a photograph a curved trail of a particle that was both too thin and too curved to be caused by a particle with the mass of a proton, but that curved off in the wrong direction to be an electron. He had discovered the positron: the electron's positively charged twin.

Anderson (1905–91) won the Nobel Prize in Physics in 1936 for his discovery of the positron, sharing the honour with Victor Hess, who discovered cosmic radiation. Anderson followed up his achievement later the same year by identifying, together with his graduate student Seth Neddermeyer, another subatomic particle, the muon, which received less recognition. Shortly afterwards, the field of experimental physics moved on, and expensive particle accelerators became the main tool in the search for further subatomic particles. Anderson, whose institution did not wish to invest in such expensive equipment, was soon left behind.

The only possible conclusion seemed to be that this track, indeed, was the track of a positively charged electron.

Between 1928 and 1931, British theoretical physicist Paul Dirac published an equation and a series of papers that moved towards theorizing the existence of a particle with the opposite charge of an electron. The equation was an early attempt to unify Einstein's theories of relativity and the classical physics of Newton. Around the same time, physicists Chung-Yao Chao and Dmitri Skobeltsyn noted what they thought were electrons that behaved in an atypical way, but neither man followed up on his observation.

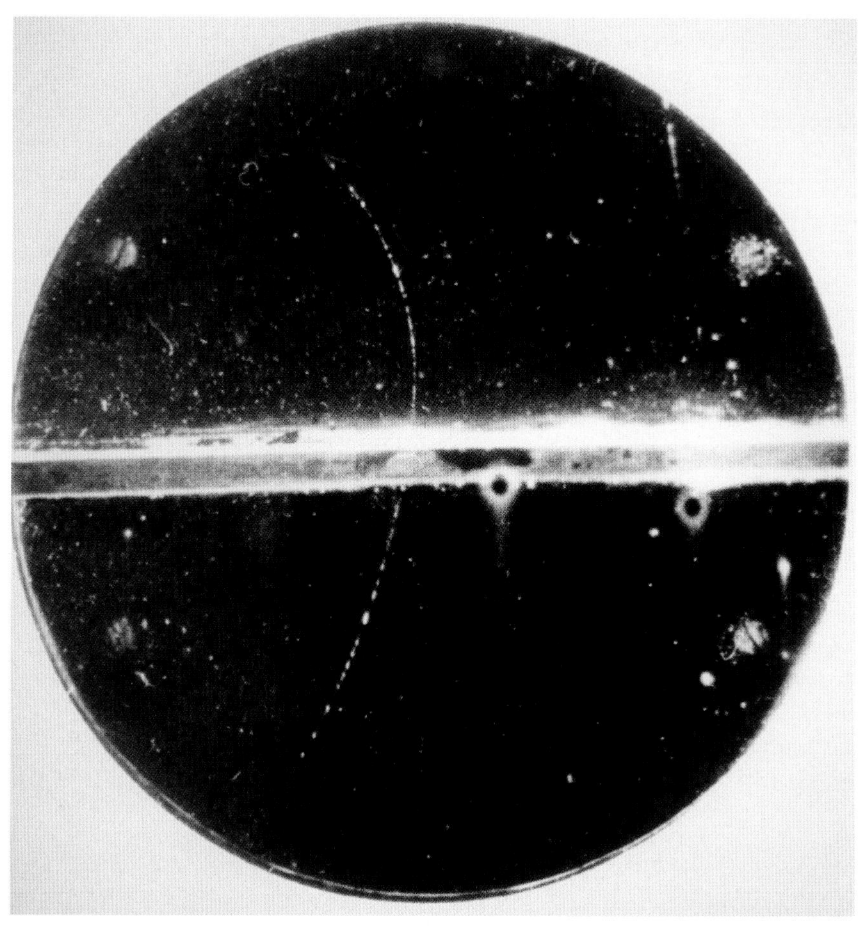

The positron was the first antimatter particle to be detected and it went a long way towards validating Dirac's equation. However, it did not solve all the problems. Theoretical physicists were soon speculating that further subatomic particles must exist, and these were subsequently detected. Before long scientists had found so many new and different types of subatomic particle that they referred to them as a 'zoo'. The next challenge was to discover how all these particles fitted together and possibly linked the evolving theories of quantum mechanics and classical physics.

This photograph from Anderson's experiment in 1932 shows the telltale ionization trail of a positron, which entered from the bottom of the cloud chamber and was slowed by the lead plate in the middle of the photograph.

After World War II, jet aircraft became faster, and the medical field struggled to keep up with the pressures that these new speeds were putting on pilots' bodies. However, the Haber brothers decided to investigate a possible means to negate, at least temporarily, gravity itself.

SIMULATING ZERO GRAVITY

FRITZ AND HEINZ HABER

1951

At the Air Force School of Aviation Medicine at Brooks Air Force Base, Texas, Fritz Haber and his brother, Heinz, theorized that a pilot in control of an aircraft that was flying in an undulating parabola would experience moments of increased and reduced G-force. They further speculated that with a sufficiently angled parabola—for example, one of 45 degrees above or below horizontal—pilots might experience a brief period of complete weightlessness lasting for as long as thirty-five seconds. Intrigued by this proposition, Air Force chiefs authorized a series of flights to assess the Habers' theory. Test pilots Chuck Yeager and Scott Crossfield followed the proposed flight plan and confirmed that weightlessness could be achieved. However, they reported that its effects lasted only twenty seconds. Through further experimentation, US pilots discovered that if they used a cargo plane, passengers as well as pilots could experience weightlessness. On an average 3,050-metre (10,000 ft) parabola, pilots would experience the absence of weight when approaching and going over the top of the peak, and they would experience around 1.8 times the force of normal gravity when going through the trough of the parabola. The experience of rapidly changing G-forces made many of them feel nauseous, which gave rise to the phrase 'vomit comet'.

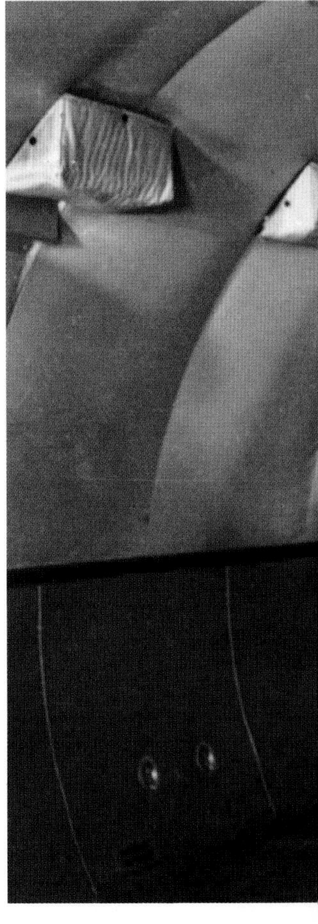

German-born Fritz (1912–98) and Heinz (1913–90) Haber were brought to the United States at the end of World War II as part of Operation Paperclip, which aimed to harness German expertise. Fritz was an engineer, while Heinz was a physicist.

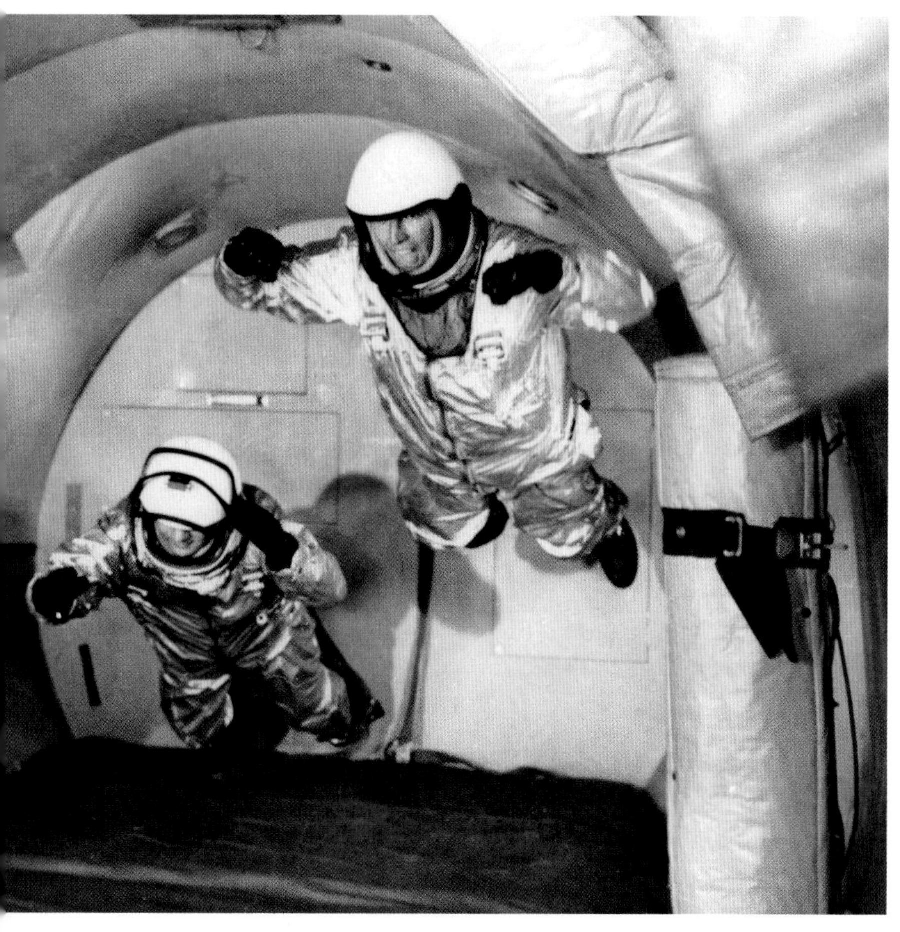

The Haber brothers' findings have proved to be an invaluable resource, and their premise is used to help those who are training for a space mission to prepare for zero gravity conditions. It also offers a cheaper alternative to expensive space flights for some microgravity experiments. More recently, commercial operators have begun to offer parabolic microgravity flights as a means for non-astronauts to experience the weightlessness of space within Earth's atmosphere. Professor Stephen Hawking is just one of the people to have taken a trip in a vomit comet for fun.

In 1959 the pilots of Project Mercury were among the first astronauts to train for zero gravity on board a specially equipped C-131 Samaritan transport plane.

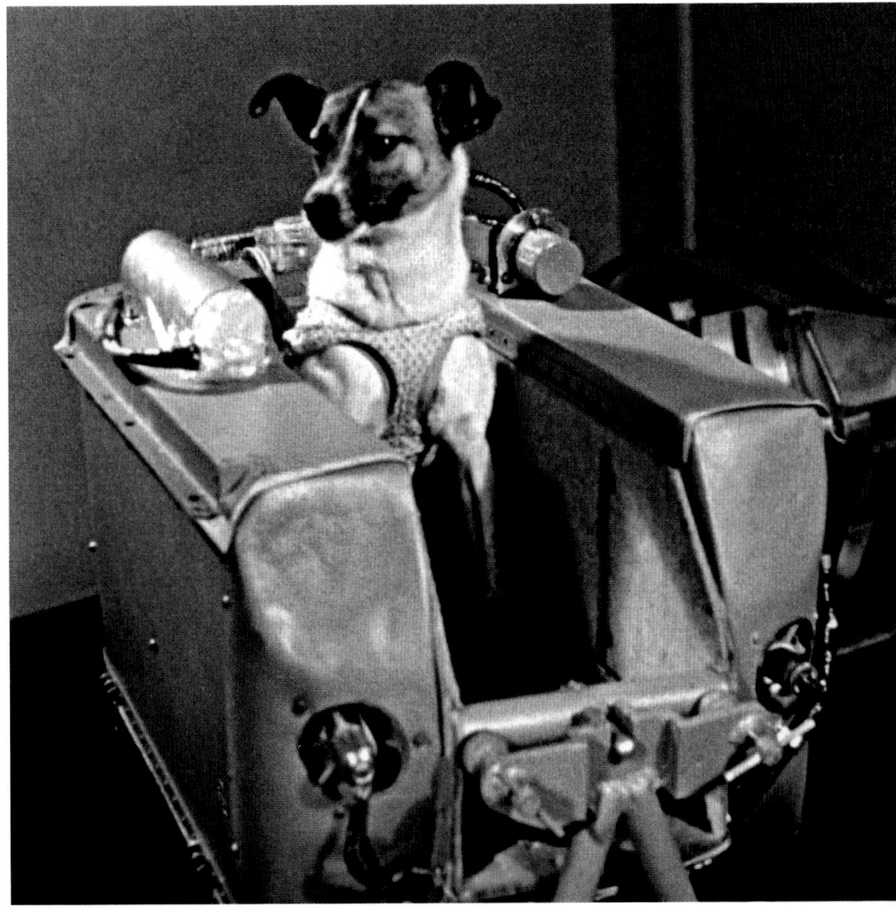

Laika was the first animal to orbit Earth and the first to die there, perishing on her fourth orbit from the heat of the cabin, where the temperature control systems failed. Subsequent missions were more successful for their animal occupants and enabled Korolyov and his team to design the re-entry and landing systems that would be essential for future manned missions. On 19 August 1960, the Russians launched Korabl-Sputnik 2, with two dogs named Belka and Strelka on board, along with 1 rabbit, 42 mice, 2 rats, flies and several plants; all were brought back to Earth alive and well.

Laika's accommodation was fitted beneath the nose cone of the Sputnik 2 rocket and she remained there for the duration of her space mission.

ANIMALS IN SPACE
SERGEY KOROLYOV
1957

Like the original Sputnik mission in October 1957, Sputnik 2 was designed and led by Soviet engineer and rocket designer Sergey Korolyov. Dogs had long been used in animal experiments in Russia and a dozen had already participated in suborbital space flights before Laika's historic mission. Like those previous pioneers, Laika was both female and a stray, selected because of her temperament and ability to cope with the long periods of inactivity that would be involved in a space flight. Commanded by Soviet leader Nikita Khrushchev to plan and execute the launch in only four weeks, Korolyov knew there was not time to develop the technologies necessary for a safe re-entry and landing: Laika's trip was planned to be only one-way. The Sputnik 2 vehicle itself was adapted from its predecessor to carry the additional payload of a dog and some scientific instruments to measure cosmic rays and solar radiation. There was also a camera and equipment on board to monitor Laika's heart rate, perspiration and blood pressure. Her accommodation for the flight was a modest padded cot in which she was secured and only able to stand or sit; she was also fitted with a spacesuit that dealt with her waste.

Having survived a lengthy spell in a brutal labour camp on a trumped-up charge during the Great Purge, Korolyov (1907–66) developed the first true intercontinental ballistic missile in the form of the R-7 Semyorka rocket. He then adapted the rocket to launch Sputnik, thereby inaugurating the space age. For security reasons, Korolyov's identity was concealed during his lifetime and he was often referred to simply as the 'chief designer'. He became a member of the Academy of Sciences of the USSR in 1958 and a crater on the far side of the moon, which was first photographed on a mission he oversaw, was named after him.

The successful launch of Sputnik, the first artificial satellite, on 4 October 1957, greatly impressed Khrushchev. It prompted him to instruct Korolyov and his team to launch a second mission with a live animal on board to coincide with the fortieth anniversary of the Soviet Revolution four weeks later. Although a science-based orbital animal flight had been planned, politics was the driving force behind the timing and the limitations of the mission.

The more time passes, the more I'm sorry about it. . . . We did not learn enough from this mission to justify the death of the dog.

OLEG GAZENKO, LAIKA'S TRAINER

Gagarin's flight marked the beginning of human space exploration. He proved that humans could survive in space. The tests performed on Gagarin when he returned to Earth also inaugurated the field of space medicine. Since then, scientists have studied the long-term effect of weightlessness on muscles and bone density, finding that both reduce over time due to the reduced load they carry in zero-g conditions. Even the hard skin on your feet peels off in space when you are not walking around.

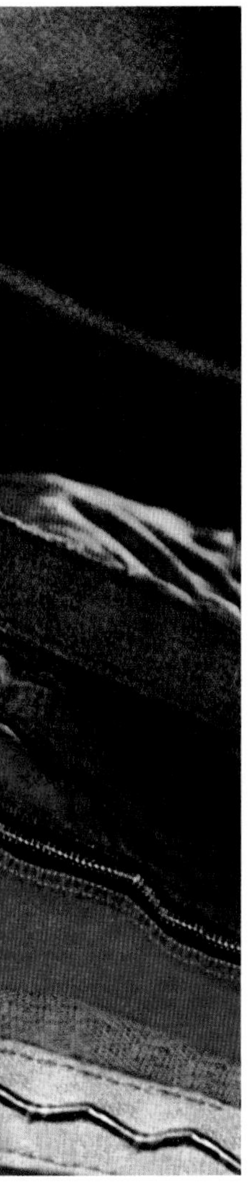

The first great goal of the space age, sought by both the United States and the Soviet Union, was a successful manned space flight. The former had been stung by the unexpected success of the first two Sputnik missions and was keen to catch up; Soviet scientists were equally determined to maintain their lead.

MANNED SPACE FLIGHT

SOVIET SPACE PROGRAMME

1961

The design of the rockets and related equipment in the early years of the Soviet space programme was the responsibility of Sergey Korolyov, but cosmonaut selection was the domain of the deputy chief of combat training for space, Nikolai Kamanin. In the weeks before the launch, Kamanin narrowed down the list of astronauts to Gherman Titov and Yury Gagarin: both were skilled pilots and both were short and therefore suited to the cramped Vostok control module. Gagarin was chosen eventually on the grounds that Titov would be a better candidate for the longer flights that were already being planned. The engineers attempted to minimize the risk of failure by automating or controlling from Earth all of the rocket's main systems, with onboard controls held under a coded lock and reserved for only emergency use. Ignoring this, Kamanin gave Gagarin the code to unlock the controls. Vostok 1 launched at 06.07 GMT on 12 April 1961, and was tracked from ground stations over the course of its single orbit. After about one-and-a-half hours, Vostok 1 began its descent, and Gagarin ejected out of the re-entry module and parachuted back to Earth as the first man to have been to space.

The letters CCCP (USSR in Russian) would prevent the misidentification of Gagarin as a US spy if he was unconscious on landing.

While not a scientist himself, Kamanin (1908–82) played a central role in the early years of manned space flight as the head of cosmonaut training in the Soviet Union from 1960 until his retirement in 1971. His experience as a pilot gave him valuable insight into the skills required by successful cosmonauts.

OPERATION FISHBOWL

US DEPARTMENT OF DEFENSE

1962

Operation Fishbowl was designed to study the effect of nuclear detonations in the upper atmosphere, the resultant electromagnetic pulses, and their impact on command and control and communications infrastructure. The nuclear devices were launched by rocket from Johnston Atoll, located about 1,400 kilometres (870 mi) south-west of Hawaii in the Pacific Ocean. Each launch was monitored from multiple stations around the Pacific—on land, sea and air—as well as by instruments on twenty-seven smaller rockets, also launched from Johnston. Of the five successful tests in the series, the largest was a 1.4-megaton hydrogen bomb dubbed Starfish Prime. It launched shortly before 11 p.m. on 8 July 1962, and detonated at an altitude of 400 kilometres (248 mi). The blast caused a brilliant aurora of high-energy particles that illuminated the sky for around seven minutes, and a larger than expected electromagnetic pulse that knocked out street lights and power lines in Hawaii. More importantly, it created an artificial radiation belt of charged electrons that caused eight orbiting satellites to fail. It was apparent that a choice would have to be made between having nuclear detonations in space and operational scientific and communications satellites.

James Van Allen (1914–2006) was the most prominent scientist involved in Operation Fishbowl. He is best known for the discovery of the radiation belts in Earth's magnetosphere that bear his name. These belts were detected by instruments designed by Van Allen and this achievement led to his recruitment for Operation Fishbowl. He was named *Time* magazine's Man of the Year in 1960 and won the Elliott Cresson Medal—the Franklin Institute's highest award—the following year.

A brilliant white flash burned through the clouds rapidly changing to an expanding green ball of irradiance.

EYEWITNESS

Although the US military had performed high-altitude nuclear tests in the late 1950s, the data from these experiments had been poor. This led scientists to design the Fishbowl tests, with more instrumentation deployed across a wide area of the Pacific. The aim was to capture better data on the electromagnetic pulse, radio blackout and aurora associated with such tests.

Following the detonation of Starfish Prime, pictured here, bright aurorae were visible widely across the Pacific region as nuclear debris interacted with the upper atmosphere.

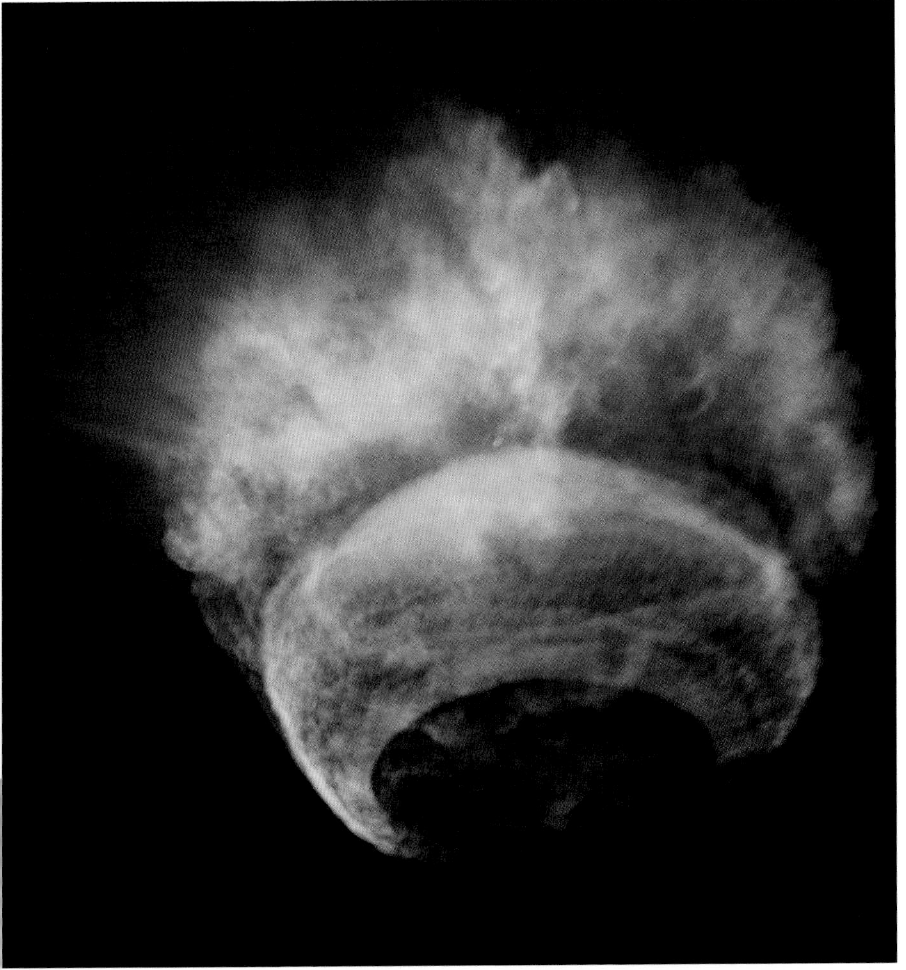

The Fishbowl tests greatly enhanced scientists' understanding of electromagnetic pulse effects created by nuclear detonations. They also led to the establishment in 1963 of the Partial Nuclear Test Ban Treaty, limiting similar nuclear detonations in the future. It prohibited its signatories—the United States, Soviet Union and United Kingdom—from atmospheric, space and underwater nuclear tests. Data from Operation Fishbowl remains valuable in the 21st century; in 2010 it was utilized in the US Defense Threat Reduction Agency report 'Collateral Damage to Satellites from an EMP Attack'.

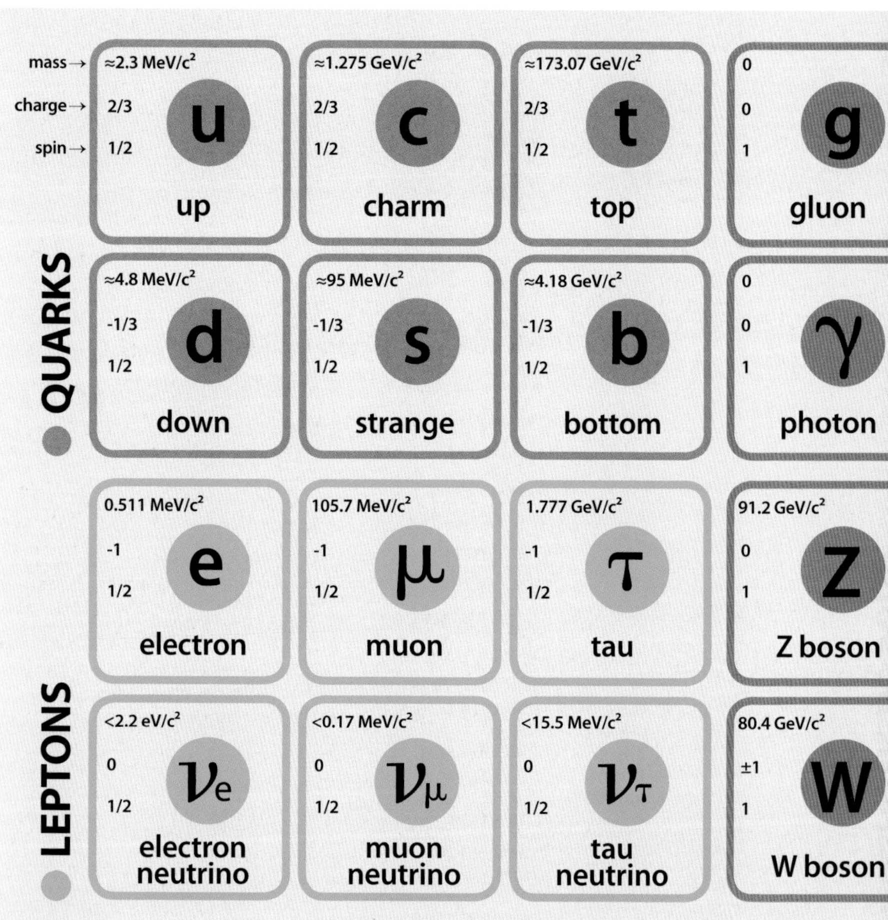

The identification of quarks led the way to the development and widespread acceptance of the standard model of fundamental particles. This model has grown to encompass six quarks—double the number proposed by Gell-Mann—six leptons (a group of particles that includes electrons and neutrinos), four force-carrying gauge bosons, and the mass-giving Higgs boson. The periodic table for particles represents a considerable simplification of many larger subatomic particles that have been identified.

Standard model particles combine to form all known matter. They are measured and defined by three fixed characteristics: mass, positive or negative charge, and quantum mechanical spin.

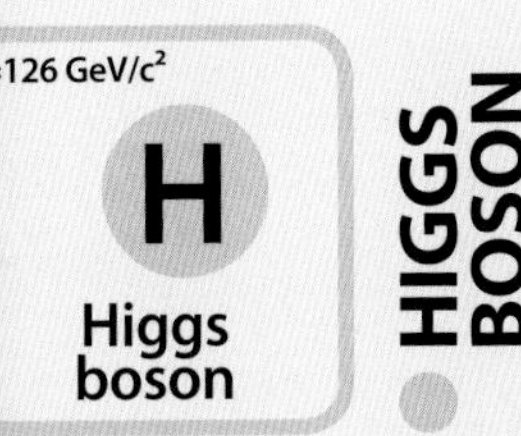

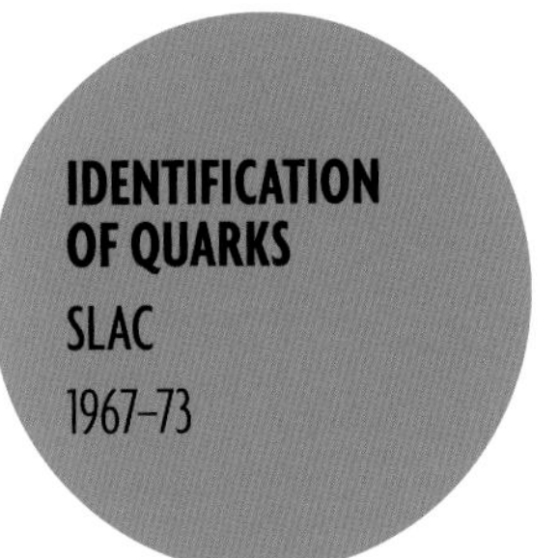

The path to creating a definitive list of the smallest particles that make up the universe began with experimental data to prove the existence of quarks. This evidence came from a series of experiments carried out at Stanford Linear Accelerator Center (SLAC) between 1967 and 1973. In these experiments, electrons, which we now understand to be fundamental particles, were fired at protons. In the first tests, the electrons bounced off the protons, but in later ones the electrons smashed the protons apart. These parts, now known as quarks, could not be detected directly, but their existence was interpreted from the change in energy of the electrons as they rebounded from the collision, in what was termed 'deep inelastic electron scattering'. Later experiments at SLAC involved enhanced detectors and firing electrons at both protons and neutrons. When these results were added to those produced by other particle acceleration experiments—involving neutrino–proton, neutrino–neutron, electron–positron and proton–proton collisions—physicists concluded that the three kinds of quarks independently proposed by US theoretical physicist Murray Gell-Mann, together with a fourth kind he had not predicted, were the best match for their observations.

By the 1960s, physicists had described so many new small particles that they had created a veritable particle zoo. This left many scientists keen to narrow down a list of the most basic fundamental particles.

While the standard model evolved from the work of many theoretical physicists, the simplification of the large number of particles to fundamental particles began with Gell-Mann (1929–) and, separately, George Zweig (1937–). Gell-Mann was awarded the Nobel Prize in Physics in 1969 for 'his contributions and discoveries concerning the classification of elementary particles and their interactions'.

The theories of relativity revolutionized physics in the early 20th century and earned German-born Einstein (1879–1955) the Nobel Prize in Physics in 1921 for 'his services to theoretical physics, and especially for his discovery of the law of the photoelectric effect'. His letter to President Roosevelt in 1939 convinced the US government to put more resources into nuclear research, paving the way for the Manhattan Project (1942–45) and the global nuclear energy industry.

Einstein is seen here in 1934 teaching in the United States. He published the theory that would become known as special relativity nearly three decades earlier in 1905.

In science, relevant experimental facts supersede theoretical arguments.

TESTING THE THEORY OF SPECIAL RELATIVITY

JOSEPH HAFELE AND RICHARD KEATING

1971

In October 1971, physicist Joseph Hafele and astronomer Richard Keating set out to test the relative time dilation effects predicted by Einstein's theory of special relativity using atomic clocks and aeroplanes. In order to give themselves a standard time at ground level, they used the atomic clocks of the US Naval Observatory in Washington, D.C. The two scientists then took two pairs of cesium beam atomic clocks, the most accurate timepieces available, on two trips around the world on commercial aeroplanes. Each clock in the pair served to check the accuracy of the other. One pair of clocks travelled eastwards, the other travelled westwards, on broadly equivalent routes. On the aeroplanes, each timepiece occupied its own seat (and was ticketed as Mr Clock). For Hafele and Keating, who travelled with the clocks, time remained constant. However, once the trips had been completed the scientists discovered that, on average, the atomic clocks travelling east had lost 59 nanoseconds (+/-10 nanoseconds) whereas the pair travelling west had gained 273 nanoseconds (+/-7 nanoseconds) when compared to the clock at the Naval Observatory. These results were broadly in line with the values predicted by Einstein's theory of special relativity, taking into account the additional gravitational time dilation effect as predicted in general relativity.

In 1905 Albert Einstein's theory of special relativity postulated that the speed of light was the same for all observers and that the laws of physics remain the same in all inertial frames of reference. One of the consequences of these ideas is that time remains relative to the observer.

The effect of Einstein's theories of special and general relativity can be noted on global positioning satellites. These satellites have internal clocks that are losing and gaining time in a similar fashion to the Hafele-Keating experiment. In order to keep their timing and geographic accuracy in tune with that on the surface of Earth, the satellites constantly correct their internal atomic clocks in line with relativity.

Mstislav Keldysh (1911–78) was one of the central architects of the Soviet space programme and he was widely regarded as its chief theoretician. His colleague, Sergey Korolyov (1907–66), was the lead rocket engineer and chief designer. Keldysh personally oversaw the inquiry into the Soyuz 11 disaster.

LEARNING FROM SPACE DISASTERS

SOVIET SPACE PROGRAMME

1971

After a successful mission to Salyut 1, the world's first space station, the crew of Soyuz 11—Georgy Dobrovolsky, Vladislav Volkov and Viktor Patsayev—began preparations to return to Earth. At an altitude of approximately 168 kilometres (104 mi), well above the boundary between Earth's atmosphere and space, six small explosive charges fired to separate the utility module from the crew module as the latter began its descent to Earth. Subsequent investigations have revealed that these charges fired simultaneously instead of consecutively, as designed. The resulting extra jolt caused two pressure valves to open prematurely to the vacuum of space. The crew members were alerted to the depressurization but there was nothing they could do. Within seconds, the small amount of air in the crew module was sucked out. Exposed to the vacuum, the cosmonauts died from asphyxiation in less than a minute. The module's automated landing procedures continued flawlessly in radio silence, and the pod set down in a remote region of Kazakhstan. Upon opening the module's hatch, the ground crew found all three cosmonauts dead.

From the outset, everyone involved in the endeavour of manned space exploration knows that it is an incredibly dangerous enterprise. A number of astronauts and cosmonauts had already died in space missions before 1971, but they had all perished within Earth's atmosphere. The Soyuz 11 disaster was the first, and so far only, time humans have died in space.

(L–R) Flight engineer Vladislav Volkov, commander Georgy Dobrovolsky and test engineer Viktor Patsayev in the cabin of the spacecraft in June 1971.

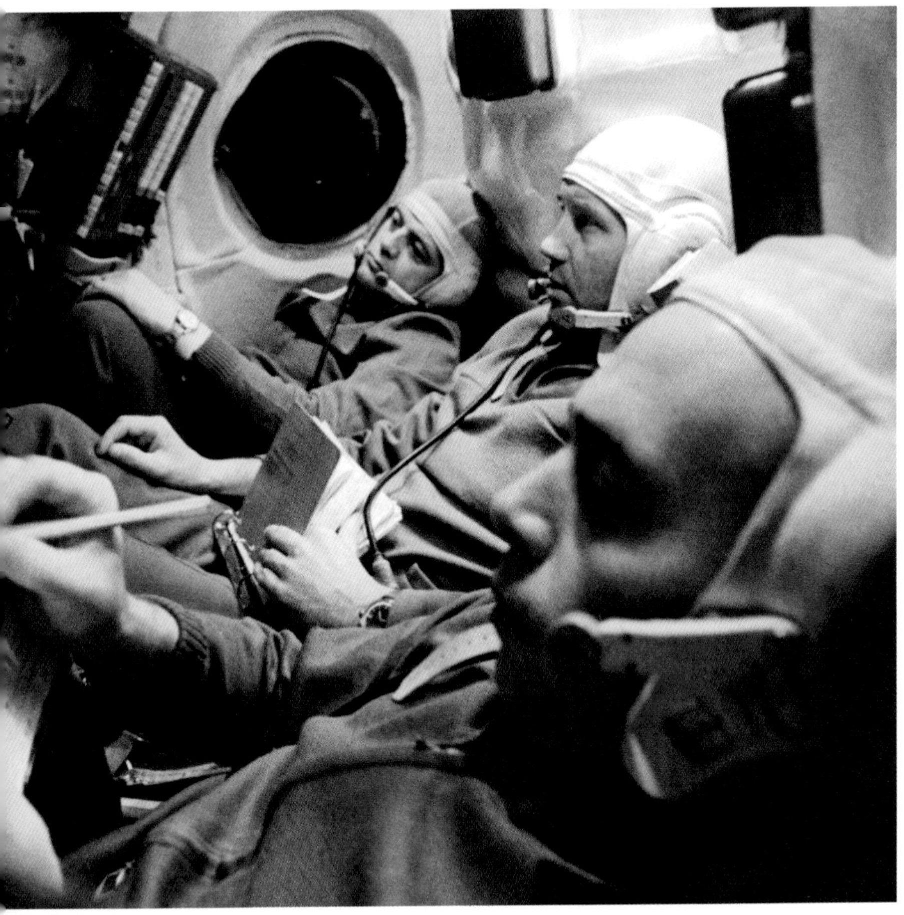

"

Outwardly, there was no damage whatsoever. [The ground crew] knocked on the side, but there was no response from within.

KERIM KERIMOV, CHAIR OF THE STATE COMMISSION

?

The crew module was redesigned extensively after the failure of the Soyuz 11 mission. The pressure valves were reconfigured to be stronger and easier to close in an emergency. Furthermore, lightweight 'rescue' spacesuits were built for cosmonauts to wear during critical phases of future missions. Known as Sokol (falcon) spacesuits, they have been designed to keep cosmonauts alive in the event of a sudden loss of pressure and remain in use on Soyuz flights today.

Having completing its main mission on 14 December 1980, Voyager 1 began the extended Voyager Interstellar Mission (VIM) a decade later, as did its sister probe Voyager 2. The aim of VIM is to chart conditions in the previously unexplored region up to and beyond the limit of the sun's magnetic field—the heliopause—and to gather data on the interstellar medium beyond. In order to help Voyager 1 maximize the remaining power from its three plutonium-based generators, NASA scientists shut down the craft's camera and any instrument no longer deemed necessary for its ongoing mission. The probe currently has only five fully operational instruments and a further two that are faulty; NASA expects to be able to keep all of these powered up to at least 2020. Data from these instruments is recorded regularly and sent back to Earth, taking more than eighteen hours to reach the Deep Space Network of antennae and the teams of scientists that monitor the craft's progress. In September 2013 NASA announced that Voyager 1 had entered interstellar space on 25 August 2012. This conclusion was based on the drop in measurements of low-energy particles from the sun, a massive increase of extra-solar cosmic rays and additional data analysis. However, some instruments are still picking up elements of the sun's influence, and it remains uncertain how far into interstellar space these last vestiges of solar wind will remain detectable.

Launched on 5 September 1977, Voyager 1 was one of two space probes to the outer planets. It was designed primarily to visit Jupiter, Saturn and some of their attendant moons. However, its mission was extended to chart the limits of the sun's influence and whatever lies beyond for as long as it has enough power for its scientific instruments to function.

As an advisor to NASA, Carl Sagan (1934–96) was responsible for the 'golden records' on board the Voyager probes. The records contain information, images and recordings about life on Earth and where in space our planet is located. They are intended for any future intelligent life forms.

If you wish to make an apple pie from scratch, you must first invent the universe.

CARL SAGAN

?

On 17 February 1998, Voyager 1 became the most distant man-made object from Earth, travelling further than the previous record holder, Pioneer 10. As of November 2015, Voyager 1 is approximately 20,013,000,000 kilometres (12,435,500 mi) from Earth, or 133 times the distance from Earth to the sun. At that time, it was still sending data back to Earth, but at some point after 2025 it will no longer have sufficient power to operate any of its instruments. When this occurs, Voyager 1's only remaining function will be as a time capsule of human civilization as it continues silently on its way through the vastness of space.

Twenty years after the VLA observed polar ice on Mercury, NASA launched the MESSENGER probe, which became the first to orbit Mercury. Its mission was to map the planet's surface, its electromagnetic field and some of its surface chemistry. As part of this mission, the probe's neutron spectrometer was used to detect the interaction of cosmic rays with the planet's surface, discovering low concentrations in the north polar region. This suggested higher quantities of hydrogen from water ice, which backed up the data from the VLA in 1991.

The radio telescopes of the VLA are deployed in an iconic Y-shaped formation.

VERY LARGE ARRAY

NATIONAL RADIO ASTRONOMY OBSERVATORY

1980–present

Built in the 1970s by the National Radio Astronomy Observatory and inaugurated in 1980, the Very Large Array (VLA) in New Mexico consists of twenty-seven individual dish antennae radio telescopes. Each one has a diameter of 25 metres (82 ft) and they are positioned along three axes of a Y-shaped track. The VLA can be placed in various configurations, thereby allowing its operators to alter its area of focus. Since its first light (the first use of a telescope), more than 2,500 scientists have employed the array in their celestial observations. For example, in 1991 the VLA was used as the receiver for what was, in effect, a giant radar experiment in which a beam of microwaves was fired at Mercury. The microwaves bounced off the planet and were received at the VLA in the same way that the array observes the microwaves emitted from distant stars. Surprisingly, unexpected high reflectivity was noted in the planet's north and south polar regions, similar to that seen on Mars. Despite the massively high temperatures on Mercury's surface, caused by its proximity to the sun, analysis of the data indicated that the temperature in shaded areas within the craters could be as low as -148.1°C (-234.6°F), easily cold enough for ice to exist.

Radio telescopes detect electromagnetic radiation emitted in space. The more that are linked together, the wider or more focused the results. This was the reasoning behind the creation of the VLA.

US radio astronomer David Heeschen (1926–2012) was among the first to determine accurately the spectra of radio galaxies, but it was his skills as an administrator that enabled him to deliver world-class facilities for radio astronomy. He is known as the 'father of the VLA', largely for his tireless work in shepherding the observatory personally and overseeing the design of many crucial elements.

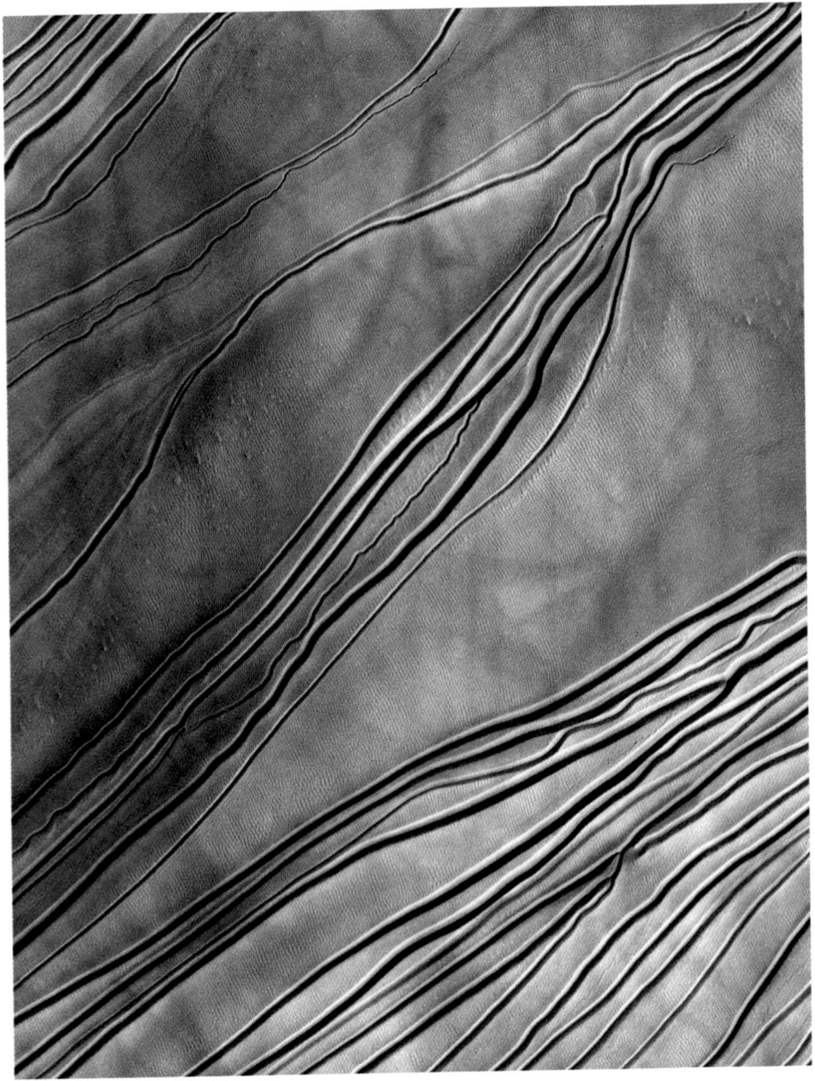

?

Data from newer and more advanced missions to Mars have served to increase our knowledge and understanding of our near neighbour. After years of searching for traces of water on Mars, NASA announced findings in 2015 indicating that liquid water exists there right now. The next question for planners of future Mars missions is whether there might still be any simple life forms on the planet.

WATER ON MARS
NASA
2002–present

The NASA Mars Odyssey spacecraft began its science mapping mission in the orbit of Mars in February 2002. It used its gamma ray spectrometer to monitor the energy of gamma rays and neutrons, emanating from the planet's surface as a result of the impact of cosmic rays. The measurements of the energy of these neutrons and gamma rays enabled the scientists to calculate which elements were present in the surface soil and to what extent. The main element that scientists were searching for was hydrogen, an essential component of water. By the end of May 2002, the orbiter had finished mapping the whole of the surface of Mars and had found significant deposits of hydrogen in the south polar region, many of which were concentrated deep in the soil. These results suggested that up to half of the deeper soil could be composed of ice, but a further lander mission was required to confirm the data. It took another six years for the Phoenix Mars lander to reach the red planet, but while it was there the craft dug up small clumps of icy dirt that sublimated in the Martian atmosphere over the course of four days. These results, and the detection of water vapour by the lander's mass spectrometer, led NASA scientists to conclude that the data from Mars Odyssey were accurate. They announced their evidence of water on Mars on 31 July 2008.

US astro-biologist and geochemist William Boynton was the lead investigator on the gamma ray spectrometer on the Mars Odyssey probe. Unlike many more recent NASA missions, Mars Odyssey has no set end date and will carry on for as long as the probe continues to function.

Ever since Italian astronomer Giovanni Schiaparelli first observed what he called the *canali*, or channels, of Mars in 1877, there has been speculation about the presence of water, a key indicator for life, on the planet. When humans began to send probes to Mars in the 1960s, one of the things they were looking for was evidence of water.

Our quest on Mars has been to "follow the water," in our search for life in the universe . . .

JOHN GRUNSFELD, ASTRONAUT

Scientists thought the gullies on this sand dune in Russell Crater on Mars were created by water ice, but recent research suggests they were formed by frozen carbon dioxide.

It is extremely difficult to get probes to Mars, and of the forty-three missions to the red planet as of November 2015, more than half have failed.

The CMB was first detected by accident in 1964 by US radio astronomers Arno Penzias and Robert Wilson, using a large microwave horn antenna. Their discovery was of great importance to cosmology and provided crucial evidence to support the Big Bang theory.

COSMIC MICROWAVE BACKGROUND

EUROPEAN SPACE AGENCY

2009–13

Launched in 2009, the European Space Agency's unmanned Planck spacecraft was the latest in a series of probes designed to measure the cosmic microwave background (CMB). This is the omnipresent leftover radiation from the time when the universe cooled sufficiently for the first atoms to form, approximately 380,000 years after the Big Bang. It is the first light in the universe and can be detected in all directions of the sky. Planck was able to observe the universe at a wider range of frequencies than ever before, thus providing the most detailed map to date. In order to observe more precisely the tiny fluctuations in the CMB, it carried two main scientific instruments: the low frequency instrument, which focused on radio signals in the 27 to 77 GHz range, and the high frequency instrument, focusing on radio and microwave signals in the 84 GHz to 1 THz range. During its mission, Planck scanned the whole of the sky twice and continued to send back data until fuel ran out in 2013.

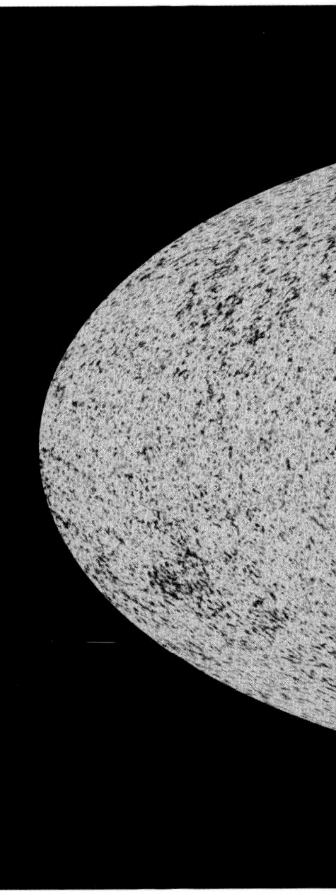

Indonesian-born Jan Tauber (1960–) served as the lead project scientist for the Planck mission from its approval in 1996. He gained his doctorate in radio astronomy in 1990 and joined the European Space Agency in 1992.

Out of such a simple object as the cosmic microwave background we can derive so much information about the origin and evolution of our very complex universe.

JAN TAUBER, ESA SCIENTIST

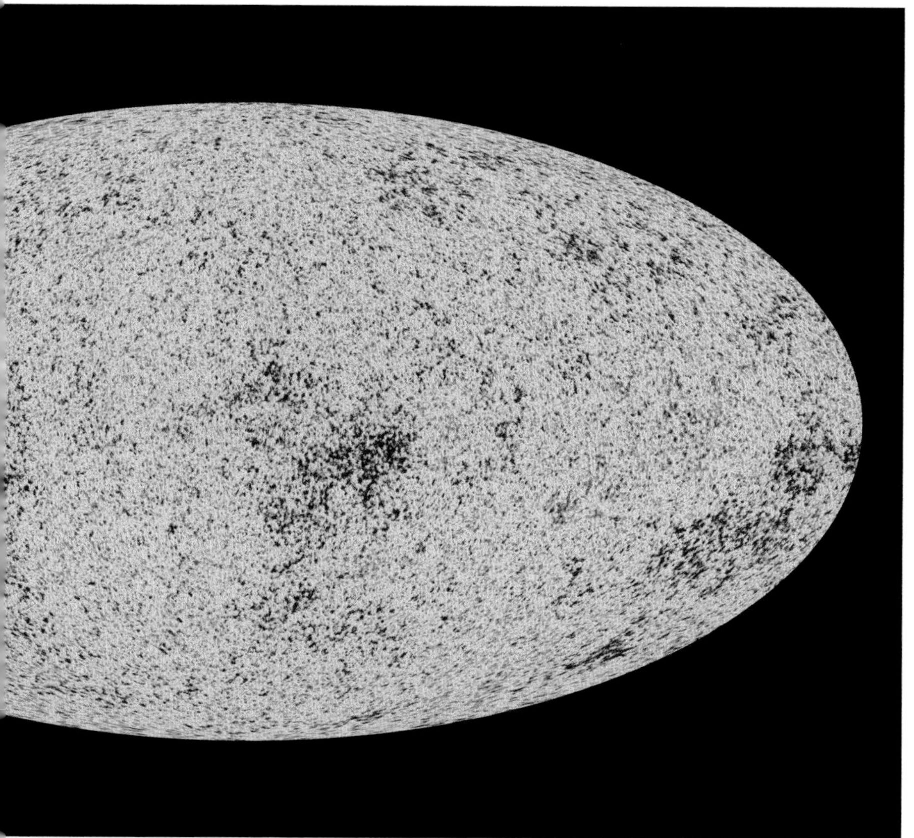

Mission scientists released data from the Planck probe in 2013 and 2015. Broadly, the results matched the predictions and results of earlier CMB missions. In addition to producing the highest resolution map of the early universe to date, it gathered data that pointed to the universe being 13.813 billion years old, slightly older than previously thought. It also confirmed the current standard model of the origins of the universe, known as the Lambda-Cold Dark Matter model. The model estimates that the visible parts of the universe—atoms, stars and planets—make up only 4.9 per cent of the universe, whereas dark matter and dark energy, about which scientists know very little, accounts for the remaining 95.1 per cent.

This Planck map of the cosmic microwave background shows the oldest light in the known universe and the seeds of all the structures, galaxies and nebulae that we see around us in the universe today.

Neutrinos are important but poorly understood, and a better understanding of their behaviour is the goal of cutting-edge physics. Takaaki Kajita and Arthur McDonald won the Nobel Prize in Physics in 2015 for discovering that neutrinos oscillate between three different forms, which shows they have some mass. The exploratory research being carried out at the IceCube Neutrino Observatory may well lead to exciting discoveries that have not yet been anticipated. As the search for neutrinos continues, the next goal is to detect one that has an energy level that is a magnitude or two higher than that of Bert or Ernie. Such a neutrino would, theoretically, have its origins in the cosmic background radiation of the universe.

The IceCube Neutrino Observatory, located near the South Pole, is the largest of its kind in the world.

The possibility of the neutrino was first suggested in 1930 by Wolfgang Pauli, one of the pioneers of quantum physics. However, it was not until 1956 that US physicists Frederick Reines and Clyde Cowan collaborated to prove its existence.

SEARCHING FOR NEUTRINOS

ICECUBE COLLABORATION

2010–present

Neutrinos are fundamental particles in the make-up of the universe. These tiny subatomic particles are all around us, but they lack any electric charge, have very little weight and rarely interact with their surroundings. Consequently, they are difficult to detect. However, when neutrinos are found they are likely to be intact and can therefore be used to decipher the distant phenomena in which they originate. The IceCube Neutrino Observatory, located close to Amundsen–Scott South Pole Station in Antarctica, was built between 2004 and 2010. It is a giant experiment that is being conducted to detect high-energy neutrinos of extra-terrestrial origin. The observatory consists of thousands of omnidirectional detector modules. These are sunk into a cubic kilometre of ice, in strings of up to sixty at a time, at depths that vary from about 1,500 to 2,500 metres (4,900–8,200 ft). The ice helps to shield the sensors from non-neutrino signals. Scientists monitor the modules for any sign of neutrinos; it took two years and a recalibration of the sensitivity of the equipment to detect the first twenty-eight neutrinos. Among these were two particularly high-energy signals, nicknamed Bert and Ernie, which were discovered in 2012 when scientists were examining IceCube data. The signals indicated a reasonable likelihood that the neutrinos that had generated them had been created somewhere outside the solar system.

Those responsible for the IceCube Neutrino Observatory are known collectively as the IceCube Collaboration. It comprises three hundred people from forty-five institutions in twelve different countries, many of whom remain actively involved in the vast experiment.

The confirmation of the existence of the Higgs boson, and its current characterization as a particle with a mass of 126 GeV, completes the standard model of fundamental particles as we currently understand them. However, it leaves many unanswered questions, such as dark matter and dark energy. Data confirming or denying the existence of further particles will likely come from future experiments at the LHC particle accelerator, which was upgraded in 2013 and 2014 to work with collisions at even higher energies.

The eight tube-like structures of ATLAS create the powerful magnetic fields that are necessary to confine all the particles created in proton collisions.

DETECTING THE HIGGS BOSON

CERN

2011–13

At 27 kilometres (17 mi) in circumference and stretching across the borders of France and Switzerland, the Large Hadron Collider (LHC) is the largest machine and science laboratory ever built. Run by the European Organization for Nuclear Research (CERN), it consists of a circular particle accelerator that has accelerated two beams of particles to more than 99 per cent of the speed of light, smashing them together next to various detectors. Between 2011 and 2013, the experimental focus of the LHC was to find whether scientists could detect the then-theoretical particle the Higgs boson by smashing together two accelerated beams of protons. The resultant collision was expected to release a wide variety of subatomic particles, but many of these, including the Higgs, decay too quickly to be detected directly. Consequently, ATLAS—the largest broad-range detector—was tasked with spotting and recording the various decay chains—termed 'decay channels'—of the numerous collision products. The data from the collisions had to be analysed for signs of the Higgs. In 2012 the initial results indicated that a particle with a mass of 126 GeV consistent with the Higgs had been observed. The result was confirmed a year later.

When physicists found that none of the fundamental particles of the standard model explained why each particle had a different mass, they concluded that mass must come from a field associated with a further particle.

British theoretical physicist Peter Higgs (1929–) is best known for postulating in 1964 the existence of the boson that now bears his name. In 2013 the Higgs particle was confirmed experimentally, and he was awarded (with François Englert) the Nobel Prize in Physics for 'the theoretical discovery of a mechanism that contributes to our understanding of the origin of mass of subatomic particles'.

SOURCES

CHAPTER 1 – THE HUMAN BODY AND MEDICINE

p. 10–11 Vaccination – Edward Jenner

Jenner, E. *An Inquiry into the Causes and Effects of the Variolae Vaccinae* (London, 1798)

Underwood, E. A. (1949) Edward Jenner, Benjamin Waterhouse and the introduction of vaccination into the United States. *Nature, 163*, 823–828

p. 12–13 Yellow Fever – Stubbins Ffirth

Ffirth, S. *A treatise on malignant fever: with an attempt to prove its non-contagious nature* (University of Pennsylvania, 1804)

Reed, W., Carroll, J., Agramonte, A. and Lazear, J. W. The Etiology of yellow fever—A Preliminary Note. Reprinted from the *Proceedings of the Twenty-eighth Annual Meeting of the American Public Health Association in Indianapolis*, October (1900)

p. 14–15 Digestion – William Beaumont

Beaumont, W. *Experiments and Observations on the Gastric Juice and the Physiology of Digestion*, (Plattsburgh, 1833)

Beaumont, W. (1825) Joseph Lovell. A case of wounded stomach. *Philadelphia Medical Recorder, 8*. 14–19

p. 16–17 Darwin's Finches – Charles Darwin

Gould, J. (1837) Remarks on a Group of Ground Finches from Mr. Darwin's Collection, with Characters of the New Species. *Proceeding of the zoological Society of London, 5*. 4–7

Lack, D. (1945) The Galapagos Finches (Geospizinae): A Study in Variation. *Occasional Papers of the California Academy of Science, 21*. 1–152

p. 18–19 Heredity – Gregor Mendel

Mendel, G. (1866) Versuche uber Pflanzenhybriden (Experiments Concerning Plant Hybrids) *Verhandlungen des naturforschenden Vereines in Brunn* (Proceedings of the Natural History Society of Brünn), *IV* (1865), 3–47

Fisher, R. A. (1936). Has Mendel's work been rediscovered? *Annals of Science I* 115–137

p. 20–21 Roundworm Life Cycle – Giovanni Grassi

Grassi, B. (1881) Noto interno ad alcuni parassiti dell'uomo III. Interno all'Ascaris lumbricoides, *Gaz. Osp. Milano. 2*(432)

Grassi, B. (1888) Weiteres zur Frage der Ascaris Entwicklung (More on the Development of Ascaris), *Centralblatt für Bakteriologie und Parasitenkunde*, 3(24). 748–9

p. 22–23 Corpse Fauna – Eduard Ritter von Niezabitowski

Niezabitowski, E. R. v. (1902) Experimentelle Beiträge zur Lehre von der Leichenfauna (Experimental contributions to the study of the corpse fauna). *Vierteljahresschrift für gerichtliche Medizin und öffentliches Sanitätswesen* (1). 44–50

p. 24–25 Protopathic and Epicritic Pain – Henry Head

Rivers, W. H. R., Head, H. (1908) A human experiment in nerve division. *Brain, 31*. 324–450

Head, H., Rivers, W. H. R., Sherren, J. (1905) The afferent nervous system from a new aspect. *Brain, 28*, 99–115

p. 28–29 Discovery of Penicillin – Alexander Fleming

Fleming, A. (1929) On the antibacterial action of cultures of a penicillium, with special reference to their use in the isolation of B. Influenzae. *British Journal of Experimental Pathology, 10*(3). 226–236

Chain, E., Florey, H. W., Gardner, A. D., Heatley, N. G., Jennings, M. A., Orr-Ewing, J., & Sanders, A. G. (1940) Penicillin as a Chemotherapeutic Agent. *The Lancet, 236*(6104). 226–228

p. 30–31 Cardiac Catheterization – Werner Forssmann

Forssmann, W. (1929) Die Sondierung des rechten Herzens (The probing of the right heart). *Klinische Wochenschrift, 8*. 2085–2087

p. 32–33 Sleep/Wake Cycles – Nathaniel Kleitman

Kleitman, N. (1923) The Effects of Prolonged Sleeplessness on Man. *American Journal of Physiology, 66*. 67–92

Kleitman, N., Aserinsky, E. (1953) Regularly Occurring Periods of Eye Motility, and Concomitant Phenomena, During Sleep. *Science, 118*. 273–274

p. 34–35 HeLa Cells – George Gey

Scherer, W. F., Syverton, J. T., Gey, G. O. (1953) Studies on the Propagation in Vitro of Poliomyelitis Viruses. IV. Viral Multiplication in a Stable Strain of Human Malignant Epithelial Cells (Strain Hela) Derived from an Epidermoid Carcinoma of the Cervix. *Journal of Experimental Medicine, 97*(5). 695–710

Skloot, R., *The Immortal Life of Henrietta Lacks*, 2010

p. 38–39 Head Transplant – Vladimir Demikhov

Demikhov, V. P. *Experimental transplantation of vital organs*. (New York, Consultants Bureau, 1962)

p. 40–41 Rapid Deceleration –John Paul Stapp

Stapp, J. P. (1951) Human tolerance to deceleration: summary of 166 runs. *Journal of Aviation Medicine, 22*(1). 42–45

Stapp, J. P. (1951) *Human exposures to linear deceleration*. United States Air Force, Wright Air Development Center

p. 42–43 Kidney Transplant – Joseph Murray

Merrill, J. P., Murray, J. E., Harrison, J. H., & Guild, W. R. (1956) Successful homotransplantation of the human kidney between identical twins. *Journal of the American Medical Association, 160*(4). 277–282.

Murray, J. E., Merrill, J. P., Harrison, J. H., Wilson, R. E., & Dammin, G. J. (1963) Prolonged survival of human-kidney homografts by immunosuppressive drug therapy. *New England Journal of Medicine, 268*(24). 1315–1323

Dammin, G. J., Couch, N. P., & Murray, J. E. (1957) Prolonged survival of skin homografts in uremic patients. *Annals of the New York Academy of Sciences, 64*(5). 967–976.

p. 44–45 Operation Whitecoat – US Army

Mole, R. L., and Mole, D. M. (1998) *For God and Country: Operation Whitecoat: 1954–1973*. TEACH Services, Incorporated

p. 46–47 Stimoceiver – José Manuel Rodríguez Delgado

Delgado, J. M. R. et al. (1964) Radioestimulation cerebral en toros de lidia. VIII Reun. *Nacional Soc. Ciencias Fisiologicas*

Delgado, J. M. R. *Evolution of the physical control of The Brain* (1965)

Delgado, J. M. R. *Physical control of the mind: Toward a psychocivilized society* (1969)

Delgado, J. M. R. Experiments with Mind Control (Remote Control Brain Implants). *February 24, 1974 edition of the Congressional Record, No. 262E, 118*

p. 48–49 Drinking Helicobacter Pylori – Barry Marshall

Warren, J. R., and Marshall, B. (1983) Unidentified curved bacilli on gastric epithelium in active chronic gastritis. *The Lancet, 321*(8336). 1273–1275

Marshall, B. J., Armstrong, J. A., McGechie, D. B., and Glancy, R. J. (1985) Attempt to fulfil Koch's postulates for pyloric Campylobacter. *The Medical Journal of Australia, 142*(8). 436–439

Peura, D. A., Pambianco, D. J., Dye, K. R., Lind, C., Frierson, H. F., Hoffman, S. R., & Marshall, B. J. (1996) Microdose 14C-urea breath test offers diagnosis of Helicobacter pylori in 10 minutes. *The American journal of gastroenterology, 91*(2), 233–238

p. 50–51 Animal Cloning – Keith Campbell and Ian Wilmut

Campbell, K.H., McWhir, J., Ritchie, W. A., and Wilmut, I. (1996) Sheep cloned by nuclear transfer from a cultured cell line. *Nature, 380*(6569). 64–66

Campbell, K. H., and Wilmut, I. (1997) Totipotency or multipotentiality of cultured cells: applications and progress. *Theriogenology, 47*(1). 63–72

p. 52–53 Caenorhabditis Elegans Genome – Richard Durbin and Jean Thierry-Mieg

Explore Worm Biology www.wormbase.org

Sequencing Consortium. (1998) Genome sequence of the nematode C. elegans: A platform for investigating biology. *Science, 282*. 2012–2018

p. 54–55 Human Genome Project – Craig Venter

National Human Genome Research Institute – www.genome.gov

lVenter, J. C. et al. (2001) The sequence of the human genome, *Science 291* (5507). 1304–1351

CHAPTER 2 – PSYCHOLOGY AND BEHAVIOUR

p. 58–59 Animal Electricity – Luigi Galvani

Galvani, L., and Aldini, G. (1792) De Viribus Electricitatis In Motu Musculari Comentarius Cum Joannis Aldini Dissertatione Et Notis; Accesserunt Epistolae ad animalis electricitatis theoriam pertinentes. *Apud Societatem Typographicam*

p. 60–61 Conditioned Reflex – Ivan Pavlov

Pavlov, I. P. (1902) *The work of the digestive glands* (Charles Griffin & Co. London, 1910)

p. 62–63 Conditioned Fear – John Watson

Watson, J. B., and Rayner, R. (1920) Conditioned emotional reactions, *Journal of Experimental Psychology, 3*(1). 1-14

p. 66–67 Neural Stimulation – Wilder Penfield

Penfield, W., Boldrey, E. (1937) Somatic motor and sensory representation in the cerebral cortex of man as studied by electrical stimulation, *Brain, 60*. 389–443

p. 68–69 Stuttering In Children – Wendell Johnson and Mary Tudor

Tudor, M. (1939) *An experimental study of the effect of evaluative labeling of speech fluency*. Unpublished master's thesis

p. 70–71 Social Conformity – Solomon Asch

Asch, S. E. (1956) Studies of independence and conformity: I. A minority of one against a unanimous majority. *Psychological monographs: General and applied, 70*(9)

p. 72–73 Group Identity – Muzafer Sherif

Sherif, M., and Sherif, C. W. *Groups in harmony and tension; an integration of studies of intergroup relations* (New York, 1953)

p. 76–77 Obedience to Authority – Stanley Milgram

Milgram, S. (1963) Behavioral study of obedience. *The Journal of Abnormal and Social Psychology, 67*(4), 371

p. 80–81 Prison Behaviour – Philip Zimbardo

Haney, C., Banks, W. C., & Zimbardo, P. G. (1973) Study of prisoners and guards in a simulated prison. *Naval Research Reviews, 9*(1-17)

p. 82–83 Diagnosis of Mental Illness – David Rosenhan

Rosenhan, D. L. (1973) On being sane in insane places. *Science, 179*(4070). 250–258

p. 86–87 Free Will – Benjamin Libet

Libet, B., Wright, E.W., Gleason, C. (1982) Readiness-potentials preceding unrestricted 'spontaneous' vs. pre-planned voluntary acts. *Electroencephalography Clinical Neurophysiology, 54*. 322–335

p. 88–89 Brain–Computer Interface – Yang Dan

Dan, Y., et al. (1999) Reconstruction of Natural Scenes from Ensemble Responses in the Lateral Geniculate Nucleus. *Journal of Neuroscience, 19*(18). 8036–8042

p. 90–91 Concentration and Perception – Dan Simmons and Christopher Chabris

Simons, D. J., & Chabris, C. F. (1999) Gorillas in our midst: Sustained inattentional blindness for dynamic events. *Perception-London, 28*(9). 1059–1074

p. 92–93 fMRI and Neuroscience – Craig Bennett

Bennett, C. M., Baird, A. A., Miller, M. B., & Wolford, G. L. (2011) Neural correlates of interspecies perspective taking in the post-mortem Atlantic salmon: an argument for proper multiple comparisons correction. *Journal of Serendipitous and Unexpected Results, 1*. 1–5

p. 94–95 Mars Mission Simulation – IBMP and ESA

Kurmazenko, E., Morukov, B., Demin, E., Kamaletdinova, G., and Khabarovskiy, N. *Life Support System Virtual Simulators for Mars-500 Ground-Based Experiment.* (INTECH Open Access Publisher, 2012)

CHAPTER 3 – SOCIETY

p. 98–99 Linnaean Taxonomy – Carl Linnaeus

Linnaeus, C. *Systema naturae* (1735)

p. 100–101 Effects of Electricity – Johann Wilhelm Ritter

Ritter, J. W. (1802) *Beyträge zur nähern Kenntniss des Galvanismus und der Resultate seiner Untersuchung*, 2

p. 102–103 Ether as a General Anaesthetic – William Morton

Bigelow, H. J. (1846) Insensibility during surgical operations produced by inhalation. *The Boston Medical and Surgical Journal, 35*(16). 379–382, 309–317

p. 104–105 Epidemiology – John Snow

Snow, J. *On the mode of communication of cholera.* (London: 1855).

p. 106–107 Discovery of Mauveine –William Perkin

Perkin, W. H. (1862–1863) On Mauve or Aniline-Purple. *Proceedings of the Royal Society of London, 12.* 713–715

Meth-Cohn, O., and Smith, M. (1994) What did W. H. Perkin actually make when he oxidised aniline to obtain mauveine? *Journal of the Chemical Society, Perkin Transactions, 1*(1). 5–7

p. 108–109 Cocaine as a Local Anaesthetic – Karl Koller

Koller, C. (1884) On the use of cocaine for producing anæsthesia on the eye. *The Lancet, 124*(3197). 990-992

p. 110–111 Magnifying Transmitter – Nikola Tesla

Tesla, N., and Marincic, A. (1978) *Colorado Springs Notes, 1899–1900*, 3

p. 112–113 Producing Ammonia – Fritz Haber

Haber, F. (2002) The synthesis of ammonia from its elements. Nobel Lecture, June 2, 1920. *Resonance, 7*(9). 86–94

p. 114–115 X-Ray Crystallography – Lawrence and William Bragg

Bragg, W. L. (1912) The Specular Reflection of X-rays. *Nature, 90*(410)

p. 116–117 Exposure to Chemical Weapons – Joseph Barcroft

Barcroft, J. (1931) The toxicity of atmospheres containing hydrocyanic acid gas. *Journal of Hygiene, 31*(1). 1–34

p. 120–121 Transoceanic Migration –Thor Heyerdahl

Heyerdahl, T., *American Indians In The Pacific. The Theory Behind The Kon-Tiki Expedition* (1952)

p. 122–123 Subliminal Advertising – James Vicary

Karremans, J. C., Stroebe, W., & Claus, J. (2006) Beyond Vicary's fantasies: The impact of subliminal priming and brand choice. *Journal of Experimental Social Psychology, 42*(6). 792–798

p. 124–125 Schmidt Pain Index – Justin Schmidt

Schmidt, J. O., Blum, M. S., and Overal, W. L. (1983) Hemolytic activities of stinging insect venoms. *Archives of Insect Biochemistry and Physiology, 1*(2). 155–160

Schmidt, J. O. (1990) Hymenoptera venoms: striving toward the ultimate defense against vertebrates. *Insect defenses: adaptive mechanisms and strategies of prey and predators.* SUNY Press. 387–419

p. 128–129 Human Cyborg – Kevin Warwick

Warwick, K., Gasson, M., Hutt, B., Goodhew, I., Kyberd, P., Andrews, B., Teddy, P., & Shad, A. (2003) The application of implant technology for cybernetic systems. *Archives of neurology, 60*(10). 1369–1373

Warwick, K., *I, cyborg* (University of Illinois Press, 2004)

p. 130–131 Transgenic animals – Nexia

Jones, J., Rothfuss, H., Steinkraus, H., & Lewis, R. (2010) Transgenic goats producing spider silk protein in their milk; behavior, protein purification and obstacles. *Transgenic Research 19*(1) 135–135

Evans, D. L., Schmidt, J. O. (1990) *Insect defenses: adaptive mechanisms and strategies of prey and predators.* SUNY Press. 353–386

p. 132–133 Folding@Home – Vijay Pande

Pande, V., and Shirts, M. S. (2000) Screen savers of the world unite! *Science, 290* 1903–1904

p. 134–135 Anti-Loitering Device – Howard Stapleton

Stapleton, H. (2011). *U.S. Patent No. 8,031,058.* Washington, D.C.: U.S. Patent and Trademark Office.

p. 136–137 Dinosaur gait – Bruno Grossi et al.

Grossi, B., Iriarte-Díaz, J., Larach, O., Canals, M., & Vásquez, R. A. (2014) Walking Like Dinosaurs: Chickens with Artificial Tails Provide Clues about Non-Avian Theropod Locomotion. *PloS one, 9*(2)

CHAPTER 4 – THE PLANET

p. 140–141 Lightning Rod - Benjamin Franklin

Franklin, B. (1754) *New Experiments and Observations on Electricity Made at Philadelphia in America.* (London, 1754)

p. 142–143 Discovery of Oxygen – Joseph Priestley

Priestley, J., and Hey, W. (1772) Observations on Different Kinds of Air. *Philosophical transactions, 62.* 147–264

p. 144–145 Periodic Table – Dmitri Mendeleev

Mendeleev, D. (1869) The relation between the properties and atomic weights of the elements. *Journal of the Russian Chemical Society, 1.* 60–77

p. 148–149 Anatomy of the Atom – Ernest Rutherford

Rutherford, E. (1911). LXXIX. The scattering of α and ß particles by matter and the structure of the atom. *The London, Edinburgh, and Dublin Philosophical Magazine and Journal of Science, 21*(125). 669-688.

p. 150–151 Radiometric Dating – Arthur Holmes

Holmes, A. (1911) The association of lead with uranium in rock-minerals, and its application to the measurement of geological time. *Proceedings of the Royal Society of London. Series A, Containing Papers of a Mathematical and Physical Character, 85*(578). 248–256

p. 152–153 Pitch Drop – Thomas Parnell

Edgeworth, R., Dalton, B. J., and Parnell, U. T. (1984). The pitch drop experiment. *European Journal of Physics, 5*(4) 198

p. 154–155 Nuclear Fission – Enrico Fermi

Fermi, E. (1946) The development of the first chain reacting pile. *Proceedings of the American Philosophical Society, 90.* 20–24

p. 158–159 Carbon-14 Dating – Willard Libby

W. F. Libby (1946) Atmospheric Helium Three and Radiocarbon from Cosmic Radiation *Physical Review, 69* (11–12). 671–672

p. 160–161 Mapping the Ocean Floor – Marie Tharp and Bruce Heezen

Heezen, B. C., Ewing, M. and Tharp, M. (1959) The Floors of the Oceans I. The North Atlantic. *Geological Society of America Special Papers, 65.* 1–126

p. 162–163 Thermonuclear Fusion – Edward Teller

Teller, E. (1955) The work of many people. *Science, 121*(3139). 267–275

p. 166–167 Chaos Theory – Edward Lorenz

Lorenz, E. N. (1963) Deterministic nonperiodic flow. *Journal of the atmospheric sciences, 20*(2). 130–141

p. 168–169 Climate Change – Willi Dansgaard and Claude Lorius

Raynaud, D., and Lorius, C. (1973) Climatic implications of total gas content in ice at Camp Century. *Nature, London*, 243. 283–284

p. 172–173 Underwater Living – Florida International University

Sebens, K. P., Bernardi, G., Patterson, M. R. and Burkepile, D. (2013) Saturation Diving and Underwater Laboratories: How Underwater Technology Has Aided Research on Coral Biology and Reef Ecology. *Research and Discovery: The Revolution of Science through Scuba. Washington DC: Smithsonian Institution, 39.* 39–52

p. 174–175 Predicting Earthquakes – EarthScope

Ellsworth, W. L., Hickman, S. H. and Zoback, M. D. *Seismology in the Source: The San Andreas Fault Observatory at Depth.* Institute of Statistical Mathematics

Zoback, M., Hickman, S., Ellsworth, W. and the SAFOD science team (2011) Scientific drilling into the San Andreas Fault Zone—An overview of SAFOD's first five years. *Sci. Drill, 11*(1). 14–28

p. 176–177 Isolating Graphene – Andre Geim and Konstantin Novoselov

Novoselov, K. S., Geim, A. K., Morozov, S. V., Jiang, D., Zhang, Y., Dubonos, S. A. and Firsov, A. A. (2004) Electric field effect in atomically thin carbon films. *Science, 306*(5696). 666–669

CHAPTER 5 – THE UNIVERSE

p. 180–181 Interference of Light – Thomas Young

Young, T. (1804) The Bakerian lecture 1803: Experiments and calculations relative to physical optics. *Philosophical transactions of the Royal Society of London, 94.* 1–16

p. 184–185 Testing for Luminiferous Ether – Albert Michelson and Edward Morley

Morley, E. W. and Michelson, A. A. (1887) On the Relative Motion of the Earth and of the Luminiferous Ether. *Sidereal Messenger, 6.* 306–310

p. 186–187 Testing the Theory of General Relativity – Arthur Eddington

Einstein, A. (1952) The foundation of the general theory of relativity. *The Principle of Relativity. Dover Books on Physics, 1.* 109–164

p. 188–189 Hubble's Law – Edwin Hubble

Hubble, E. (1929) A relation between distance and radial velocity among extra-galactic nebulae. *Proceedings of the National Academy of Sciences, 15*(3). 168–173

p. 196–197 Manned Space Flight – Soviet Space Programme

Vico, L. et al. (2000) Effects of long-term microgravity exposure on cancellous and cortical weight-bearing bones of cosmonauts. *The Lancet, 355*(9215). 1607–1611

p. 200–201 Identification of Quarks – SLAC

Riordan, E. M. (1992) The discovery of quarks. *Science, 256.* 1287–1293

p. 202–203 Testing the Theory of Special Relativity – Joseph Hafele and Richard Keating

Hafele, J. C., and Keating, R. E. (1972) Around-the-world atomic clocks: predicted relativistic time gains. *Science, 177*(4044). 166–168

p. 206–207 Voyager Probes – NASA

Gurnett, D. A., Kurth, W. S., Burlaga, L. F. and Ness, N. F. (2013) In situ observations of interstellar plasma with Voyager 1. *Science, 341*(6153). 1489–1492

p. 208–209 Very Large Array – National Radio Astronomy Observatory

Slade, M. A., Butler, B. J., & Muhleman, D. O. (1992) Mercury radar imaging: Evidence for polar ice. *Science, 258*(5082), 635–640

p. 210–211 Water On Mars – NASA

Evans, L. G., Reedy, R. C., Starr, R. D., Kerry, K. E., and Boynton, W. V. (2006) Analysis of gamma ray spectra measured by Mars Odyssey. *Journal of Geophysical Research: Planets 111*(E3)

p. 212–213 Cosmic Microwave Background – European Space Agency

Planck Collaboration, I. Planck 2015 results. I. Overview of products and results. 2015.

p. 214–215 Searching for Neutrinos – IceCube Collaboration

Collaboration, S. R. (2013). Recent Highlights from IceCube. 33rd International Cosmic Ray Conference, Rio De Janeiro 2013 – The Astroparticle Physics Conference.

p. 216–217 Detecting the Higgs Boson – CERN

Higgs, P. W. (1964) Broken symmetries and the masses of gauge bosons. *Physical Review Letters, 13*(16). 508

INDEX

Page numbers in **bold** refer to illustrations

PICTURE CREDITS

Every effort has been made to trace all copyright owners but if any have been inadvertently overlooked, the publishers would be pleased to make the necessary arrangements at the first opportunity.

Cover image: © ian nolan / Alamy Stock Photo
p.2: Magnifing Transmitter, Nikola Tesla (see p. 110)

8-9 NATIONAL INSTITUTES OF HEALTH/SCIENCE PHOTO LIBRARY 10 SHEILA TERRY/SCIENCE PHOTO LIBRARY 13 KING'S COLLEGE LONDON/SCIENCE PHOTO LIBRARY 14-15 © Corbis 16-17 PAUL D STEWART/SCIENCE PHOTO LIBRARY 19 SHEILA TERRY/SCIENCE PHOTO LIBRARY 20 KING'S COLLEGE LONDON/SCIENCE PHOTO LIBRARY 22-23 John B. Carnett/Bonnier Corporation via Getty Images 24-25 Reproduced from W. H. R. Rivers and Henry Head. A Human Experiment in Nerve Division. Brain (1908) 31 (3): 323-450 (Fig. 5.) By permission of Oxford University Press on behalf of The Guarantors of Brain 26 Wellcome Library, London 29 Photo by SSPL/Getty Images 30 Filsoufi F, Carpentier A. www.themitralvalve.org 32-33 Photograph courtesy of Mammoth Cave National Park, Park Museum Collections 34-35 NATIONAL INSTITUTES OF HEALTH/SCIENCE PHOTO LIBRARY 36 SCIENCE SOURCE/SCIENCE PHOTO LIBRARY 38-39 SPUTNIK/SCIENCE PHOTO LIBRARY 40-41 © US Air Force Photo / Alamy Stock Photo 42-43 Brigham and Women's Hospital Archives 46-47 AP Photo/Charlie Neibergall 48 SCIENCE PHOTO LIBRARY 50-51 PHILIPPE PLAILLY/SCIENCE PHOTO LIBRARY 53 SINCLAIR STAMMERS/SCIENCE PHOTO LIBRARY 54-55 © Martin Shields / Alamy Stock Photo 56-57 By permission of Alexandra Milgram 58-59 SCIENCE SOURCE/SCIENCE PHOTO LIBRARY 60-61 Sovfoto/UIG via Getty Images 62-63 The Drs. Nicholas and Dorothy Cummings Center for the History of Psychology, The University of Akron 67 © INTERFOTO / Alamy Stock Photo 68-69 AP Photo 71 © Tom Howey 72-73 The Drs. Nicholas and Dorothy Cummings Center for the History of Psychology, The University of Akron 74 SCIENCE SOURCE/SCIENCE PHOTO LIBRARY 76-77 By permission of Alexandra Milgram 78-79 Nina Leen/The LIFE Picture Collection/ Getty Images 80 © Duke Downey/San Francisco Chronicle/San Francisco Chronicle/Corbis 83 Wikimedia Commons: Center building at Saint Elizabeths, August 23, 2006, User: Tomf688, https://commons.wikimedia.org/wiki/File:Center_building_at_Saint_Elizabeths,_August_23,_2006.jpg, https://creativecommons.org/licenses/by-sa/2.5/deed.en for CC BY-SA 2.5 84-85 SUSAN KUKLIN/SCIENCE PHOTO LIBRARY 87 ZEPHYR/SCIENCE PHOTO LIBRARY 88-89 Figure 2B. Stanley, G.B., Li, F.F. and Dan, Y., 1999. Reconstruction of natural scenes from ensemble responses in the lateral geniculate nucleus. The Journal of Neuroscience, 19(18), pp.8036-8042. 90-91 Figure provided by Daniel Simons. From Simons, D. J., & Chabris, C. F. (1999). Gorillas in our midst: Sustained inattentional blindness for dynamic events. Perception, 28, pp.1059-1074. 92-93 Craig Bennett, http://prefrontal.org 94-95 NATALIA KOLESNIKOVA/AFP/Getty Images 96-97 © Corbis 98 MIDDLE TEMPLE LIBRARY/SCIENCE PHOTO LIBRARY 101 SCIENCE SOURCE/SCIENCE PHOTO LIBRARY 102-103 © Corbis 105 BRITISH LIBRARY / SCIENCE PHOTO LIBRARY 106 SSPL/Getty Images 108 Courtesy of Robert E. Greenspan, MD, www.CollectMedicalAntiques.com 110-111 © Everett Collection Historical / Alamy Stock Photo 112-113 © Deutsches Museum, Munich, Archive, CD68715 115 © World History Archive / Alamy Stock Photo 116 LIBRARY OF CONGRESS/SCIENCE PHOTO LIBRARY 119 Image courtesy of the National Archives at Atlanta 120-121 Image courtesy of the Kon-Tiki Museum, Oslo 122-123 Walter Daran/ The LIFE Images Collection/Getty Images 124-125 Image courtesy of Brian Fisher, www.AntWeb.org 126 Mark Kauffman/The LIFE Picture Collection/Getty Images 128-129 JAMES KING-HOLMES/SCIENCE PHOTO LIBRARY 130-131 Field Test Independent Film Corps 132-133 Image courtesy of Folding@Home: http://folding.stanford.edu 134-135 Jonathan Player/REX/Shutterstock 136-137 Image courtesy of Jose Iriarte-Diaz and Bruno Grossi 138-139 © Kip Evans Photography 140 © Lebrecht Music and Arts Photo Library / Alamy Stock Photo 142-143 Oxford Science Archive/Print Collector/Getty Images 145 SPUTNIK/SCIENCE PHOTO LIBRARY 146-147 Keystone-France\Gamma-Rapho via Getty Images 148 PROF. PETER FOWLER/SCIENCE PHOTO LIBRARY 151 NATURAL HISTORY MUSEUM, LONDON/SCIENCE PHOTO LIBRARY 152 Wikimedia Commons: Pitch Drop Experiment with John Mainstone, John Mainstone, The University of Queensland, https://commons.wikimedia.org/wiki/File:Pitch_drop_experiment_with_John_Mainstone.jpg, https://creativecommons.org/licenses/by-sa/3.0/deed.en for CC BY-SA 3.0 154-155 SSPL/Getty Images 157 t GUSTOIMAGES/SCIENCE PHOTO LIBRARY 157 b GUSTOIMAGES/SCIENCE PHOTO LIBRARY 158 Hansel Mieth/The LIFE Picture Collection/Getty Images 161 NOAA/SCIENCE PHOTO LIBRARY 162-163 U.S. DEPT. OF ENERGY/SCIENCE PHOTO LIBRARY 164-165 PhotoQuest/Getty Images 166-167 SCOTT CAMAZINE/SCIENCE PHOTO LIBRARY 169 ROYAL INSTITUTION OF GREAT BRITAIN / SCIENCE PHOTO LIBRARY 170 LAWRENCE LAWRY/SCIENCE PHOTO LIBRARY 172-173 © Kip Evans Photography 174-175 UC REGENTS, NATL. INFORMATION SERVICE FOR EARTHQUAKE ENGINEERING/SCIENCE PHOTO LIBRARY 177 Massimo Brega, THE LIGHTHOUSE/SCIENCE PHOTO LIBRARY 178-179 MAXIMILIEN BRICE, CERN/SCIENCE PHOTO LIBRARY 181 ROYAL INSTITUTION OF GREAT BRITAIN / SCIENCE PHOTO LIBRARY 182-183 Lambert/Getty Images 184-185 Universal History Archive/UIG/Getty Images 187 ROYAL ASTRONOMICAL SOCIETY/SCIENCE PHOTO LIBRARY 188 Boyer/Roger Viollet/Getty Images 191 CARL ANDERSON/ SCIENCE PHOTO LIBRARY 192-193 NASA 194 © SPUTNIK / Alamy Stock Photo 196-197 SPUTNIK/SCIENCE PHOTO LIBRARY 200-201 Wikimedia Commons: Standard model of elementary particles: the 12 fundamental fermions and 4 fundamental bosons, Own work by uploader, PBS NOVA, Fermilab, Office of Science, United States Department of Energy, Particle Data Group, https://commons.wikimedia.org/wiki/File:Standard_Model_of_Elementary_Particles.svg, https://creativecommons.org/licenses/by/3.0/deed.en for CC BY 3.0 203 SSPL/Getty Images 204-205 Sovfoto/UIG via Getty Images 206-207 NASA/JPL-Caltech/SCIENCE PHOTO LIBRARY 208-209 MyLoupe/UIG via Getty Images 210 NASA/JPL/UNIVERSITY OF ARIZONA/SCIENCE PHOTO LIBRARY 212-213 EUROPEAN SPACE AGENCY, The Planck Collaboration/SCIENCE PHOTO LIBRARY 214-215 © Xinhua / Alamy Stock Photo 216-217 MAXIMILIEN BRICE, CERN/SCIENCE PHOTO LIBRARY

Thanks to Jane Laing and Ruth Patrick for commissioning this book and to Becky Gee for her tireless work editing it. Thanks also to my wife Mel Johnson, who offered support and encouragement throughout. Finally, I'd like to acknowledge various experts whose answers to my questions were invaluable: Greg Anderson and Kaye Shedlock at EarthScope, Craig Bennett, Daniel Nye Griffiths, Thomas Potts at Aquarius Reef Base, Howard Stapleton at Compound Security Solutions and all the librarians at UCLA.

Dedicated to my Nan, Betty Bond, and my nieces and nephews Ren and Una Cave and Evan, Lauren, Naomi, Olivia and Owen Johnson, each in their own ways little scientists.

First published in the United Kingdom in 2016 by
Thames & Hudson Ltd, 181A High Holborn,
London WC1V 7QX

This book was designed and produced by
Quintessence Editions Ltd.
The Old Brewery, 6 Blundell Street,
London, N7 9BH

Editor	Becky Gee
Designer	Tom Howey
Picture Researcher	Hannah Phillips
Production Manager	Anna Pauletti
Editorial Director	Ruth Patrick
Publisher	Philip Cooper

British Library Cataloguing-in-Publication Data
A catalogue record for this book is available from
the British Library

ISBN 978-0-500-292006

Printed in China

To find out about all our publications, please visit
www.thamesandhudson.com
There you can subscribe to our e-newsletter, browse or download our current catalogue, and buy any titles that are in print.